Springer Series in
Electronics and Photonics 27

Edited by Walter Engl

Springer Series in Electronics and Photonics

Editors: D. H. Auston   W. Engl   T. Sugano

Managing Editor: H. K. V. Lotsch

This series was originally published under the title
Springer Series in Electrophysics
and has been renamed starting with Volume 22.

Volumes 1–20 are listed on the back inside cover

L. Treitinger
M. Miura-Mattausch (Eds.)

Ultra-Fast Silicon Bipolar Technology

With 125 Figures

Springer-Verlag Berlin Heidelberg New York
London Paris Tokyo

Dr. Ludwig Treitinger
Dr. Mitiko Miura-Mattausch
SIEMENS AG, Central Research and Development,
Otto-Hahn-Ring 6, D-8000 München 83, Fed. Rep. of Germany

Series Editors:

Dr. David H. Auston

Columbia University, Dept. of Electrical Engineering, New York, NY 10027, USA

Professor Dr. Walter Engl

Institut für Theoretische Elektrotechnik, Rhein.-Westf. Technische Hochschule,
Templergraben 55, D-5100 Aachen, Fed. Rep. of Germany

Professor Takuo Sugano

Department of Electronic Engineering, The Faculty of Engineering,
The University of Tokyo, 7-3-1, Hongo, Bunkyo-ku, Tokyo, 113, Japan

Managing Editor: Dr. Helmut K. V. Lotsch

Springer-Verlag, Tiergartenstraße 17
D-6900 Heidelberg, Fed. Rep. of Germany

ISBN 3-540-50638-1 Springer-Verlag Berlin Heidelberg New York
ISBN 0-387-50638-1 Springer-Verlag New York Berlin Heidelberg

The text was word-processed using PS™ software.

Printing: Druckhaus Beltz, 6944 Hemsbach/Bergstr.
Binding: J. Schäffer GmbH & Co. KG, 6718 Grünstadt
2154/3150-543210 – Printed on acid-free paper

Preface

Since the first bipolar transistor was investigated in 1947, enormous efforts have been devoted to semiconductor devices. The strong worldwide competition in fabricating metal-oxide-semiconductor field-effect transistor (MOSFET) memories has accelerated the pace of developments in semiconductor technology. Bipolar transistors play a major role due to their high-speed performance. Delay times of about 20 ps per gate have already been achieved. Because of this rapid technological progress, it is difficult to predict the future with any certainty. In 1987 a special session on ultrafast bipolar transistors was held at the European Solid-State Device Research Conference. Its aim was to summarize the most recent developments and to discuss the future of bipolar transistors. This book is based on that session but also includes contributions by other participants, such that a broad range of up-to-date information is presented. Several conclusions can be drawn from this information: the first and most important is the very large potential for future progress still existing in this field. This progress is characterized by the drive towards higher speed and lower power consumption required for complex single-chip systems, as well as by several concrete technological implementations for fulfilling these demands. The second conclusion is that a large part of this potential can be realized by rather unsophisticated techniques and configurations well suited to uncomplicated transfer to fabrication. A third possible conclusion is that the design concepts underlying present devices are limited for a variety of reasons, but that alternative concepts are already being developed, e.g. heterojunctions instead of homojunctions, so that technical progress may well go on for a long time. In preparing this book, the objective was both to make it useful for the experts and simultaneously to provide a guide for newcomers to the bipolar field. All the contributors hope that this book will give a valuable overview of developments in bipolar technology and also provide an outline of future prospects in this field.

Munich, October 1988
L. Treitinger
M. Miura-Mattausch

Contents

1. History, Present Trends, and Scaling of Silicon Bipolar Technology

L. Treitinger and M. Miura-Mattausch

SIEMENS AG, Central Research and Development
D-8000 München, Fed. Rep. Germany

The bipolar transistor was first investigated at Bell Laboratories in 1947. W. Shockley, J. Bardeen, and W. Brattain were honoured by a Nobel Prize in Physics for their epochal invention. Since then, bipolar transistors have been widely used for many purposes. Applications can be divided into two broad categories: amplification and switching. As an amplifier more than a factor of 100 in amplification is available and a switching speed in the sub-ns range can be obtained.

Two pn junctions constitute a bipolar structure. The minority carrier concentration of a reverse-biased pn junction is under control of a bias applied to a nearby forward-biased junction. Figure 1.1 shows the configuration of a npn transistor as an example. Electrons are injected from the emitter to the base under a forward bias between the emitter and base. These electrons are swept further into the collector through a depletion region at the base/collector junction. This is observed as a collector current I_C. Figure 1.2a shows experimental values of I_C plotted on a logarithmic scale as a function of the base-emitter bias voltage V_{EB}. Experimentally, the collector current is exponentially related to V_{EB} [$I \propto \exp(qV/nkT)$; $n \simeq 1$]. Therefore the driving capability illustrated by $\Delta I/\Delta V$ is one current decade per 60mV at room temperature. This extremely high capability is maintained over more than 5 decades. In addition to I_C, a current flow between the base and the emitter, mainly hole injection from the base to the emitter, is also observed. The ratio of I_C to this base current I_B gives the current gain $\beta = I_C/I_B$ (Fig.1.2b), which is usually very large. The large value of β establishes

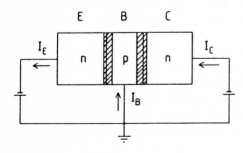

Fig.1.1. Principle of the npn bipolar transistor under a forward bias. Shaded zones represent depletion regions

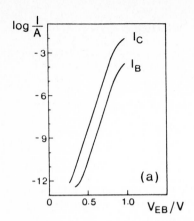

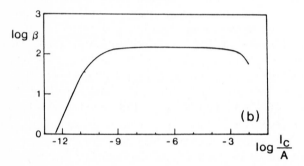

Fig.1.2a,b. Typical input and transfer characteristics of npn transistors (emitter mask size of 2×8 μm^2): **(a)** current-voltage characteristic, **(b)** current gain β

the efficiency of the bipolar transistor as an amplifier. A characteristic of the bipolar transistor is that the transistor action is caused by injection of electrons into the base region. Thus the performance of transistor switching is governed mainly by the transport of electrons in the base. Therefore a reduction of the base junction width is required for high-speed circuits so that the loss of electrons by recombination in the base region is minimized. To realize very narrow junctions a diffusion process is applied. The impurity dopant is diffused from the diffusion source into the silicon substrate. As a result, the vertical configuration of the bipolar transistor shown in Fig.1.3 is advantageous and is now adopted for common usage. To increase the electron sweeping into the collector, the doping concentration in the collector is kept very low. Usually a lightly doped epitaxial layer is used. A buried layer of heavily doped silicon is added below this layer to maintain a high-conductivity collector region.

The good reproducibility of bipolar transistors brings extremely small offsets (<1mV). Together with the large driving capability and the linearity over several decades, the small offset of bipolar transistors ensures the leading position of bipolar technology in the market of high-precision analogue circuits. The high-speed performance of bipo-

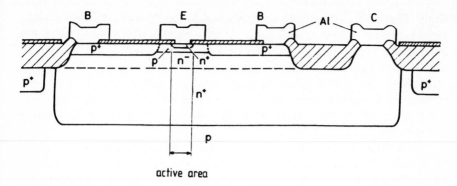

active area

Fig.1.3. Configuration of the conventional bipolar transistor: Standard Buried Collector (SBC)

lar transistors in switching, guarantees an advantage for high-speed digital circuits. To date, delay times of 30 ps have been reached for ECL (emitter-coupled logic) circuits. Gate arrays with more than 10K complexity can already be produced. Possibilities for high density and high speed RAMs have also been developed. Results show that 64K RAMs with just a few ns access time are possible.

1.1 Introduction

The last ten years have brought a rapid progress of bipolar technology, leading to a great number of new transistor configurations with surprisingly good results. The driving forces are distinctly coming from CMOS for more than one reason:

i) Modern CMOS technology contains advanced processing which can be applied to bipolar technology with minor variations at the most.

ii) The performance of CMOS circuits, especially the gate delay of unloaded gates, has reached values of about 100 ps or even less. These values are typical for ECL gates. Such CMOS innovation pushes bipolar technology to much higher switching speed in order to guarantee the survival of this rather complicated technology.

iii) The inability of CMOS circuits to drive large capacitive loads (wiring and fan out) at high speed places demands for an adequate bipolar technology. As a consequence, the main characteristics of the two technologies, supplementing one another, lead to the development of merged technologies.

The process steps taken from CMOS technology are mainly thin-film deposition techniques as well as dry-etching of the deposited

3

films. These two techniques in combination enable self-aligning process sub-sequences. These techniques are very readily controllable, both for conducting films, like polysilicon or silicides, as well as for dielectrics like silicon oxide and nitride. As an important advantage for bipolar technology, these and other process steps are taken from CMOS processing in a mature state, so that the main effort is not the development of single process steps, but the mosaic-like work of combining single steps in a successful manner.

The critical dimensions of the vertical npn transistor are not given by a lateral length as in the case of MOS transistors, but by the vertical doping profile. Therefore the lithography is not so important for bipolar devices as for MOS devices. The most critical dimensions of the vertical npn device for transistor performance are the width of the active base region and the emitter depth, both resulting from well-controlled crystal-growth techniques like deposition and diffusion. This fact enables these dimensions to be adjusted to the order of 100 nm and even below, which is smaller than by lithographic means at present.

The conventional bipolar device, usually denoted as the "standard buried collector" (SBC) is seen in Fig.1.3 [1.1-4]. The active transistor region formed by implantation steps is very small in comparison with the overall device dimensions. This means that the behaviour of the device, especially the transient performance, is dominated by the very large parasitic capacitances and resistances of the external contact regions. However, the parasitic regions cannot be reduced because of limitations due to the alignment accuracy of the lithography used.

The conventional device suffers from various characteristic drawbacks:

i) Large parasitic capacitances, i.e. large C_{BC} (base collector) and C_{CS} (collector substrate). Therefore the power consumption of a gate needed to reach the minimum of the gate delay time is high.

ii) Large extrinsic base resistance. The sheet resistance of the extrinsic base is in the range from about 300 Ω to more than 1 kΩ [1.5]. For a "short" emitter stripe, e.g. the length of the emitter, L_E, is about twice the emitter width W_E, the geometry of the external base for a device with a single-base contact is about a square of L_E. This means that the external base resistance is of the same order of magnitude or larger than the internal component of the base resistance.

iii) The vertical scaling is limited. When the base width is reduced in order to shorten the base transit time τ_F (or to increase the transit frequency f_T), the emitter depth has to be reduced by the same amount for reasons of controllability and reproducibility of junction depths. If the emitter junction depth is reduced to about 200 nm or

4

less, the emitter becomes "transparent", because the minority-carrier diffusion-length exceeds the emitter junction depth and minority carriers reach the aluminium contact. This effect strongly reduces the current gain of the device [1.6]. A SBC production technology can therefore not utilize the values of junction depth near this limit.

iv) For high-density applications like fast SRAM's, the SBC device area, especially the collector-substrate area, is too large. This is not always true for gate arrays. In this case, the integration density (and the performance) may be mainly given by the interconnect scheme, that is the multilevel wiring system [1.7,8]. A small device area is, in any event, advantageous for circuit design and performance.

In spite of these handicaps, reasonably good results have been obtained by using SBC technology [1.9-11].

Several basic changes of the transistor configuration are necessary in order to overcome the disadvantages described in i) to iv) above. The so-called self-aligned polysilicon technology is the result of this development [1.12-15]:

1) Further vertical scalability is achieved by use of a thin polysilicon film as the emitter contact. This polysilicon film also serves as the source for the diffusion of implanted n-dopants, usually arsenic, into the monosilicon region yielding very shallow and steep n^+ emitter doping profiles with emitter depths smaller than 100 nm. These shallow junctions allow for base widths of the same order of magnitude. The corresponding transit frequencies reach values as high as about 25 GHz [1.16].

2) In order to improve the unfavourable ratio between the active transistor area and the inactive base area, self-aligned emitter-base configurations have been introduced. With the same lithography as the conventional one, the area needed for contacting the active base in the monosilicon region is strongly reduced (Fig.1.4). As the main consequence, the external base resistance as well as the capacitance of the base collector junction C_{BC} are reduced. The main part of the external base contact is composed of a second thin polysilicon film doped with p-dopants, usually boron, on a thick silicon oxide separating the transistors. The sheet resistance of the external base can be kept as low as about 50 Ω by using the appropriate thickness, deposition conditions, and doping concentration of the polysilicon film [1.17-19].

3) Further reduction of the external base resistance, which is one of the most important parameters for high-speed operation, can be achieved by introducing a much more highly conducting layer in place of the polysilicon film. This very highly conductive layer can be realized by a sandwich of polysilicon and a silicide [1-20,21] making possible sheet resistances of only a few Ohms [1.22].

5

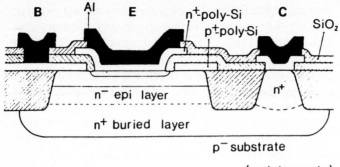

Fig.1.4. Configuration of the self-aligned polysilicon-emitter bipolar transistor

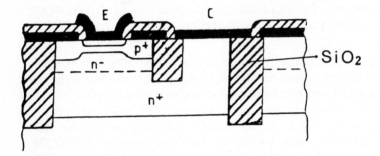

Fig.1.5. Schematic figure of the deep trench isolation

4) For lower power consumption, parasitic capacitances have to be reduced. This can be done in two different ways: By a lower doping concentration of the epitaxial layer, C_{BC} can be decreased. In this way, the device can only be operated at small collector current densities without entering the high-current regime because of the enlargement of the effective base width into the collector. Therefore this method is restricted to a certain number of applications. The second possibility is to cut down the areas of the collector-base junction and of the collector-substrate junction. This is achieved by a sort of advanced isolation scheme, preferably by a deep trench or U-groove isolation shown in Fig.1.5 as an example [1.23-29]. This second possibility yields, in addition, a strong reduction of the device area, but needs sophisticated and expensive processing. Which of the two possibilities is more appropriate depends on the application.

With all these improvements, very high performance can be achieved. In the following, the impact of each improvement on the device

6

behaviour is characterized. In Sect.1.2 we will start with the polysilicon emitter contact. Section 1.3 is devoted to the self-alignment with polysilicon layer and Sect. 1.4 to scaling problems. In Sect.1.5 present and future heterojunction transistors are demonstrated, and topical modelling problems are discussed in Sect.1.6. Section 1.7 will give an overall view of demonstration circuits, and finally, a summary and outlook are given in Sect.1.8.

1.2 Polysilicon Emitter Contact

A lot of experimental and theoretical work has been devoted to the device physics and the preparation of polysilicon contacted bipolar transistors [1.14-16, 30-39]. The main advantage of the polysilicon layer may be to maintain the separation between the very sensitive monosilicon region and the metallic contact preventing sintering effects and spikes [1.12]. The fact that the properties of the emitter contact must not be dependent on the thickness of the polysilicon layer for thicknesses larger than about 50 nm is an additional advantage for the fabrication process [1.31, 34]. At the same time, the polysilicon layer is used as the diffusion source for the emitter. Arsenic is implanted into the polysilicon, and is diffused afterwards into the monosilicon by heating. This process allows very shallow emitters (depth below 50 nm) to be obtained.

However, this new polysilicon technology also exhibits new phenomena that are impossible to understand within the conventional models:

- It is known that the current gain can be much higher for the polysilicon emitter contact than for the conventional metal contact [1.13].
- A significant tunnelling contribution is sometimes observed in the base current (Fig.1.6) [1.30, 33].

The phenomena have been explained either by low carrier mobility in the polysilicon or by the existence of an interfacial layer [1.30, 31, 33, 34]. During the process of depositing the polysilicon on the monosilicon substrate, a thin layer of contamination, which is often referred to as a "native" oxide involved in thermal oxidation or a chemical interface treatment, forms on the monosilicon surface. The thickness of the layer is normally less than a few nanometers. It has been shown that the native oxide would act as a tunnelling barrier to hole injection into the polysilicon in npn bipolar transistors. An oxide layer thickness of preferably 2 to 3 nm has been shown to result in a remarkable increase in the emitter efficiency and also in a small posi-

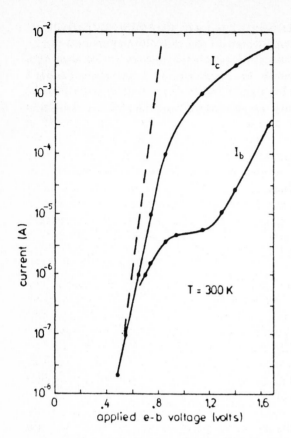

Fig.1.6. Input and transfer characteristics of transistor with thick interface oxide layer [1.30]

tive or even a negative temperature coefficient of the current gain, apparent evidence for the tunnelling current.

The polysilicon layer as a diffusion source causes a segregation at the poly-Si/mono-Si interface. It has been shown that these segregations also contribute to a reduction of the minority carrier concentration [1.34, 36]. The grain boundaries, which influence the carrier transport in the polysilicon layer, are very much dependent on the conditions of the deposition and on the subsequent annealing procedures. High-temperature annealing may cause the interface to become discontinuous and the polysilicon to regrow epitaxially [1.34, 37]. TEM (transmission electron microscopy) pictures clearly show that high-temperature annealing can break the interface oxide layer. Additionally, the character of the native oxide is dependent on the surface preparation prior to polysilicon deposition. Thus, the carrier transport properties of the polysilicon contact depend very much on the deposition conditions, impurity concentration, preparation of the interface, and annealing procedures.

For devices with a native oxide layer of appropriate thickness and/or the segregation of dopants at the polysilicon/monosilicon interface, the base current is mainly determined by the recombination and blocking of minority carriers at the interface. This has been modelled as a reduced surface recombination velocity for minority carriers. The low minority-carrier mobility in the polysilicon, caused by scattering due to the lattice disorder at the grain boundaries or segregation at boundaries, would also retard the transport of injected minority carriers. A number of such competing mechanisms for reducing the base current are modelled and lumped together in the "effective" surface recombination velocity. Values of this effective recombination velocity for the polysilicon contact are found to be about 5×10^4 cm/s and lie between the limits of a metal contact and an insulating "contact" [1.35, 38]. The tunnelling current expression for such devices has been also studied to estimate the barrier height and width caused by the oxide layer [1.35, 38].

The increase of current gain in comparison with the metal-contacted transistor was found to be a factor of two to several decades depending on the interfacial conditions. The so-called "super beta transistors", however, are of only very small practical interest owing to the very poor reproducibility and the excessive height of the emitter series resistance [1.37]. Because of its large bandgap, the insulating oxide layer forms a potential barrier not only to the minority carriers but also to the majority carriers, causing an increase of the emitter resistance. If the interface oxide thickness is larger than 1 nm, especially for small emitters, the resulting series resistance severely limits the speed and transconductance of the device [1.39]. Since the emitter current is usually at least 100 microampere, and the voltage drop due to an emitter series resistance should be limited to several millivolts, the emitter resistance has to be smaller than about 100 Ω for an emitter contact area of 1 μm^2. Therefore only polysilicon emitters with very thin or broken interfacial oxide layers are of practical interest. The interfacial oxide layer can be used to prevent the epitaxial realignment of the polysilicon layer by annealing, but has to be as thin as necessary to impede the disturbance of both the diffusion of dopants from the polysilicon layer into the monosilicon region and the electrical transport across the interface.

1.3 Self-alignment with Polysilicon Layers

The last section was devoted to the use of polysilicon to improve the emitter efficiency and to allow for further vertical scaling. This verti-

9

cal scaling leads to very shallow junctions, small base widths as well as small emitter depths [1.40,41]. In this way, high transit frequencies larger than 10 GHz and up to 30 GHz can be obtained [1.16]. The performance of circuits, e.g. the switching speed of an ECL gate, does not increase in the same manner as the transit frequency because the gate delay depends mainly on the capacitances at low currents, and at higher currents, at which the minimum delay values are found, the base resistance is of the same importance as the base transit time. The base resistance itself is composed of the internal base resistance, the effective resistance of the active base region, and the external base resistance, i.e. the sum of outer series resistances including the metal contacts. The internal base resistance is given simply by the sheet resistance of the active base and its geometry. The sheet resistance of the active base region is fixed by the demands of avoiding punch-through and of conserving the necessary current gain. Therefore, the value of this sheet resistance ("pinch resistance") cannot be far away from about 10 kΩ. The actual value of the internal base resistance is determined by the geometry, i.e. by the emitter width depending primarily on the lithography used and on the special features of the self-alignment technique. For example, the oxide spacer technique [1.14,15] leaves the effective emitter contact widths smaller than the lithographic minimum by about double the spacer width. Thus about 0.5 μm is left as the effective emitter width with a 1μm lithography, resulting in about 50 Ω for the internal base resistance of a 0.5 x 5 μm^2 transistor.

The composition of the external base resistance is much more complicated. The contact layer can be a polysilicon or a silicide layer surrounding the internal base area entirely [1.42,43]. A highly boron doped polysilicon layer with a sheet resistance between about 50 Ω and 250 Ω depending on the deposition temperature, the thickness and the doping concentration [1.17-19], is contacted by one metal contact for short emitter stripes and by two symmetric metal contacts for long emitter stripes [1.44,45]. A typical value of the external base resistance for a 0.5 x 5 μm^2 transistor is about 33 Ω for a single metal contact and a polysilicon layer of 50 Ω sheet resistance.

For silicide-layer contacting schemes, the external base resistance can be much lower (e.g. 3Ω for the 0.5x5 μm^2 transistor) according to the much lower sheet resistance of silicides of about 5 Ω for usual thickness. These values of the external base resistance are only valid in the case of very small link resistance between the internal and the external base regions. Whether or not a remarkable link resistance will appear is strongly dependent on the sequence of the process steps of the various self-alignment techniques [1.46,47]. In certain cases, the value of this link resistance can be dominant, not only for the external base resistance itself. It can also greatly increase the gate delay.

10

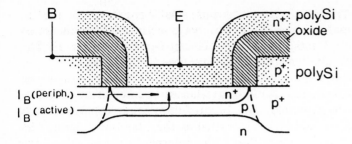

Fig.1.7. Schematic figure of the base current flow

The most important effect of the self-alignment for the emitter-base complex is cutting down the external base area in the monosilicon region leading to much smaller values of C_{BC} (Table 1.1). Together with the decrease of the base resistance, it forms the main impact of the self-alignment technique on the gate performance enhancing the maximum speed at a smaller power delay product.

Table 1.1. Comparison of the transitormparameters for Polysilicon Self-Aligned (PSA) and Standard Buried Collector (SBC). (Emitter mask size: 2 x 20 μm^2)

	PSA	SBC
Effective emitter size	1.4x19.4 μm^2	2x20 μm^2
Base resistance $R_b = R_{b,i} \pm R_{b,ex}$	60 Ω + 8 Ω	87 Ω + 46 Ω
Emitter-base capacitance C_{EB}	114 fF	196 fF
Collector-base capacitance C_{BC}	70 fF	181 fF
Collector-substance capacitance C_{CS}	140 fF	304 fF
Base transit time τ_{F0}	8 ps	13 ps

1.4 Scaling Problems

The recent development of polysilicon contacts has allowed for the shallow-junction devices and the further reduction of the extrinsic part of the bipolar device. This leads to an efficiency of the horizontal shrinkage owing to the development of lithography. The trend to further reduce the emitter size, especially the emitter stripe width in order to get a smaller intrinsic base resistance, results in an increase of the periphery-to-area ratio of the emitter. As a consequence, the influence of the emitter periphery on the transistor action increases, which may cause limitations of the conventional transistor performance, and thus limitations of the conventional scaling rules.

Because of the complicated device structure and the large parasitic contributions, scaling rules for bipolar transistors are not so simple as those for the MOS structure. For bipolar transistors, voltages usually stay out of scaling since they are already near to lower limits. The scaling rules are mainly established according to the operating current and power dissipation needs [1.40,41]. As the result of proper scaling all parameters, especially the resistances and capacitances, of the device have to be scaled in proportion, so that an appropriate circuit performance is achieved. This is done by scaling the dimensions as well as choosing the doping concentrations in a coordinated manner. Therefore considerable reduction of the lateral dimensions by using an advanced lithography technique, or by using new self-alignment schemes, has to be accompanied by the reduction of the vertical dimensions and vice versa. Thus there is a trend towards smaller emitter stripes, smaller emitter areas and hence higher current densities and higher periphery-to-area ratios.

It is known that the high peripheral contribution to the base current results in a reduction of the current gain with reduced emitter sizes with conventional transistors (Fig.1.7) [1.48,49]. However, for self-aligned transistors with polysilicon emitter contact, the current gain β can increase with reduced emitter size (Fig.1.8) [1.50]. The emitter contact area, which determines the base current, is usually smaller than the effective emitter area, which determines the collector current. The relative difference is enhanced for reduced emitter sizes and, hence, an increase of β can be observed. Thus the reduction of the horizontal dimensions can be achieved without sacrificing the current gain at least up to 1 μm lithography. For transistors with a thick

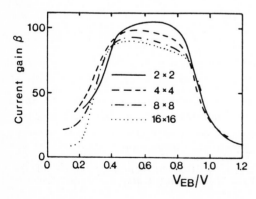

Fig.1.8. Current gain β vs voltage applied between emitter and base. Four emitter mask sizes [μm^2] are depicted

12

native oxide layer the peripheral contribution is enhanced [1.51,52]. Thus the reduction of β is prevented by keeping the interfacial region clean. If the vertical dimensions are not correctly scaled with respect to the horizontal dimensions, the reduction of current gain can also be seen even for a lithography of 1 μm [1.53].

As the vertical dimensions of the transistor are reduced, the base doping concentration must be increased to prevent the emitter–collector punch-through [1.54]. However, the base sheet resistance must be large enough to give the transistor sufficient current gain. This current gain requirement defines the value of the intrinsic base sheet resistance to be about 10 kΩ. Since the advantage of the polysilicon contact is a higher current gain, the base doping level can be increased further to reduce the intrinsic base resistance without sacrificing the current gain of the device. This may be one of the reasons why a minimum ECL gate delay as low as 34 ps has been reported for polysilicon self-aligned structures [1.16]

To prevent the base stretching effect, called the "Kirk effect" [1.55], the collector doping should be increased in accordance with the increase of the base doping concentration. However, the reduction of the breakdown voltage between the emitter and the collector will prevent the further increase of the concentration. An optimization with respect to speed, power dissipation, and breakdown voltage BV_{ECO} is required.

As the emitter size is scaled down, the influence of the periphery on the transistor performance is increased. Therefore particular attention is paid to the sidewall region, in which three doping concentrations, those of emitter, extrinsic base, and intrinsic base are merged together (Fig.1.9). The dopant concentration in the sidewall diode underneath the spacer is determined by the lateral diffusion from the base contact and the emitter polysilicon, dependent on the doping concentrations in the polysilicon layer and on the processing involved. A scaling down of the spacer width or the proximity of the extrinsic base

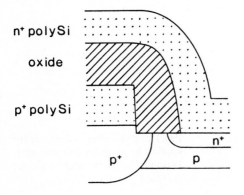

n$^+$ polySi

oxide

p$^+$ polySi

n$^+$

p$^+$ p

Fig.1.9. Schematic figure of the sidewall region

profile to the emitter profile may cause an instability of the sidewall region. A strong tunnelling at reverse as well as at forward bias occurs in the base current which induces the degradation of diodes [1.56,57]. Thus the scaling of the sidewall region is tightly limited.

The most important effect of the self-alignment technology is cutting down the parasitic external base contact region drastically. As a consequence, it reduces the extrinsic base resistance to a value much less than the intrinsic base resistance, so that the device performance is mainly governed by the active transistor part. However, the collector area is not efficiently reduced because of the large parasitic junction with the external base. To overcome this problem a new sophisticated self-aligned structure called the "sidewall-base-contact structure" (SICOS), shown schematically in Fig.1.10, has been proposed [1.58-60]. Because of the buried oxide structure underneath the extrinsic base region, the base-collector junction capacitance, which causes the most serious delay at small current densities, is reduced to the minimum. Additionally, the ratio of reverse to forward current gain is greatly improved in comparison with the conventional transistors.

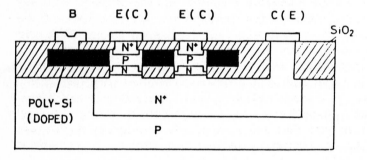

Fig.1.10 Schematic figure of the sidewall-base-contact structure (SICOS)

1.5 Heterojunction Transistors

As previously described, advanced bipolar transistors often include three features; self-alignment with poly-Si layers to reduce the external base resistance together with C_{CB}, poly-Si emitter contacts as a diffusion source to produce shallow emitters, and trench or U-groove isolation to reduce parasitic capacitances. The poly-Si emitter exhibits high emitter efficiency, mainly because of the existence of a very thin native oxide layer, sometimes partly broken, at the mono/poly-Si interface. The oxide layer causes, at the same time, high emitter resistance, which is one of the reasons to degrade the high performance of

14

such bipolar transistors. Therefore a trade-off must be made between the emitter efficiency and the emitter series resistance. The described developments are accompanied with a reduction of the base width. In order to avoid the punchthrough effect and to maintain a reasonably small base sheet resistance in the active region, the dopant concentration of the active base is enlarged in accordance with the reduction of the base width. As a result, the current gain is reduced. To overcome these problems, hetero-junction transistors with a wide-gap emitter or a narrow-gap base are now under development.

The hetero-junction introduces an extra barrier mainly for the valence bands as seen in Fig.1.11 [1.61,62]. Thus the hole injection from base to emitter is suppressed, and, as a result, an increase of efficiency in the emitter injection can be achieved. Therefore a much higher current gain can be obtained without increasing the emitter resistance. In addition to the increase of current gain, the heterojunction structure reduces the base resistance owing to the possibility of increasing the base doping further. Such hetero-structures have been intensively studied, mainly for III-V compounds [1.63-65]. However, advances in self-alignment technology provide strong motivation for the incorporation of the heterojunction into silicon transistors in a way that is compatible with existing silicon technology. In spite of the many advantages of a narrow-gap base [1.66], it is the wide gap emitter heterojunction that has mainly been studied up to now. As wide-gap heterojunctions, several kinds of materials have been investigated. Among them hydrogenated amorphous silicon a-Si:H [1.67], β-SiC [1.68-71], and microcrystalline Si [1.72] seem to be candidates. These materials can be formed at low temperatures if proper process methods are used, and thus undesired effects on impurity profiles can be avoided. The first material is deposited by discharge decomposition

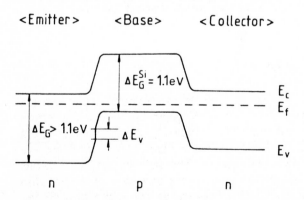

Fig.1.11. Energy band diagrams for heterojunction transistors. ΔE_v represents an extra hole barrier

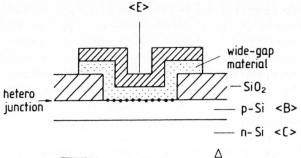

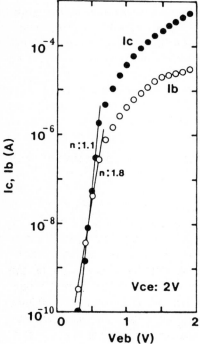

Fig.1.12. Typical transistor construction of heterojunction bipolar transistor

Fig.1.13. Collector and base currents of a heterojunction bipolar transistor as a function of the forward emitter-base voltages [1.70]

of silane, and the last two are grown by the plasma CVD method. Film thicknesses are about 100-200 nm. The band gaps are 1.7, 2.2, and 1.4 eV, respectively. A typical transistor construction is shown in Fig.1.12. The possibility of using Si-MBE for growing single crystalline oxygen-doped Si films [1.73] has also been studied. In the near future more choice will be available for the wide-gap materials. With Si-Ge alloys, narrow bandgap base heterojunction bipolar transistors have been demonstrated [1.61, 74-79].

The current gain β of hetero-junction emitter transistors has been reported to be larger than that of homojunction transistors, but it is still much lower than expected, especially at low current levels. The

16

recombination in the base current seems to lower the current gain. The value of the ideality factor [n in exp(qV/nkT)] of the collector current is around 1.1. On the other hand, the base current reaches n = 1.8 or more (Fig.1.13). The recombination current may be associated with the interface level at the heterojunction, or the crystal imperfection of the wide-gap material, or the crystal defects induced during the formation of the hetero-junction. Additionally, it is difficult to locate the p-n junction precisely at the heterojunction, if the emitter doping is performed after formation. In spite of the many problems still to be solved, the idea of the heterojunction seems to be promising. We should not forget that the advantage of the heterojunction for high-speed performance becomes dominant when all possible external parasitic capacitances and resistances are diminished. If this can be achieved, wide-gap heterojunction transistors will allow ultra-high f_T and very small gate delays at room temperature.

1.6 Topical Modelling Problems

The progress of technology as well as the need for faster circuits with lower power consumption for many applications have led to the development of a variety of high speed logic and analogue circuits. For the design and optimization of these circuits, precise modelling of the transistors' dc and ac characteristics is a prerequisite. This precise modelling has to take into account all effects arising from new transistor configurations as well as those arising from down-scaling of devices by the use of fine-line patterning. A detailed description of the basic physical phenomena in bipolar transistors and the related model can be found in the literature [1.1,79,80]. In this section, problems causing discrepancies between the existing models and real behaviour of advanced transistors are specified and characterized for some special cases.

In the following we list a selection of such modelling problems which seem to be of increasing importance for the future.

- Transit frequency and transit time, especially for higher current densities.

- Area-periphery partitioning especially for self-aligned devices.

- Sidewall effects.

- High-doping effects.

- Series resistances in the base contact, as well as in the emitter and in the collector contacts.

- Accurate methods for measuring and/or extracting the data needed for the network analysis.

17

As the transit frequency f_T of transistors exceeds about 10 GHz the resulting transit time $\tau_F = 1/(2\pi f_T)$ becomes more and more a balanced sum of contributions from the emitter, the base (width) and the collector region [1.81]. This implies that careful examination of all these contributions is required not only for process monitoring but also for a meaningful device optimization. As a consequence of the decreasing line widths obtained with advanced lithography, the transistor area decreases and therefore the current density increases for a fixed driving current. The transit frequency f_T has its maximum value for fixed bias V_{BC} at a certain current density related to the "effective" emitter area, $j_C = I_C/A_e < 1$ mA/μm^2. With further increase in the current density, the transit time τ thus increases remarkably. An exact knowledge of the causes for this $\tau(j_C)$ characteristic is still lacking.

In nearly all physical investigations of the behaviour of advanced transistor configurations the question which part of the effect observed is caused by the periphery and which by the area arises. The mask size is a measure relevant only for transistors with very large area. For small transistor sizes various factors prevent a clear relation between (emitter) mask size, electrically effective transistor area and peripheral length. These factors include the tolerances of the subsequences of self-alignment, lateral diffusion of dopants etc. A microscopic measurement of the actual geometry of the emitter contact is also not really relevant. These situations introduce a complexity to area-periphery partitioning. Additionally, the partitioning has to be performed separately for all currents because there is no reason why the effective area for the collector current should be the same as for the base current. Separate partitioning must also be done in various ranges of frequency because there is no reason why the distribution of currents should be the same for dc and for high-frequency ac currents.

Sidewall effects are found in nearly all characteristics of self-aligned transistors with narrow emitter stripes. In fact they often dominate in emitter-base breakdown characteristics. A major influence of the sidewall may be seen in the current gain (Sect.1.4, Scaling Problems) as well as in $\tau_F(j_C)$, two of the most important characteristics of the transistor. As far as possible, the sidewall effects should be suppressed by careful tailoring of the doping densities in the peripheral region. The lack of exact knowledge about these peripheral doping profiles may be the most important obstacle for the modelling. None of the methods suitable for determining the vertical doping profiles, such as SIMS or capacitance-voltage measurements, can be applied. Therefore, the sidewall doping profiles taken as the basis of 2D simulations are only assumptions or at best the results from process modelling calculations supported by single electrical results, e.g. from breakdown characteristics [1.82].

18

High-doping effects can be important for almost all parts of the transistor for various reasons. On the one hand, a knowledge of bandgap narrowing is necessary for an exact modelling of the transistor action itself. On the other hand, high-doping effects in the base may lead to remarkable contributions of forward tunnelling currents [1.56,57]. This may limit the vertical scaling which requires an increase of the base doping with decreasing base width. For future bipolar devices this could be the main reason for the substitution of homojunctions by heterojunctions. Besides this, the requirement that circuits should operate below fixed and well-known breakdown voltages also demands the inclusion of high-doping effects for exact modelling [1.81].

Series resistances, especially in the base and in the emitter contacts, are to be kept within rather small limits of tolerance, otherwise the overall performance decreases considerably. In the case of the base series resistance, an additional relatively high contribution arises, especially in the case of more sophisticated types of self-alignment like SICOS [1.58-60]. This additional series resistance is the so-called "link" resistance between the internal base region and the external base region diffused out of the p^+ polysilicon layer forming the base contact. Because of its nature, it is impossible to control the link resistance in a reproducible manner. Additionally, there is no method suitable for calculating or measuring such link resistances. The only reliable method at present is to compare the transient behaviour of transistors with and without link resistance, e.g. by suitable ring-oscillator measurements, and to estimate the "effective" link resistance by means of network analysis. Other methods, like derivation from S-parameter measurements [1.45] or from noise measurements [1.83,84] are not applicable. The lack of an exact knowledge of the doping concentrations involved, or of the resulting sheet resistance in the corresponding region, prevent the resistance from being calculated by means of geometrical and resistive data. A somewhat similar situation is found in the case of high emitter series resistances. In some cases the series resistance arising from interfacial layers can be estimated quite accurately for non-broken thin oxide layers in polysilicon emitters. In this case tunnelling calculations may give good results [1.38]. A reliable method for measuring such (current-dependent) emitter series resistances is still lacking.

Last, but not least, it should be emphasized that imperfections of network modelling as well as of device modelling are often not, or not only, consequences of imperfections of the models. In many cases the methods of measuring and of extracting the parameters may cause considerable differences between measured and simulated dc and ac characteristics. Therefore, improvement of modelling should always include simultaneously the continuous verification of parameter measurements and of the suitability of the special test configurations used.

1.7 Demonstration Circuits

High-speed silicon bipolar circuits are widely applied to all types of high-speed systems, such as main frame computers, communication systems or measuring instruments. Demonstration circuits should provide the information needed for the development of systems. Therefore various types of demonstration circuits are used, partly also prototypes of products.

In general, the basic information about the performance of a technology is given by data on various ring oscillators. With a distinct circuit technique, mainly CML or ECL because of the application fields mentioned, power delay diagrams (or collector current delay diagrams) for unloaded and loaded gates at constant voltage swing are obtained. Noteworthy results are the minimum of the power delay product found for minimum size transistors at smaller current densities, and also the minimum of the delay time found for long transistor stripes because of the benefits of lower base resistance.

A comparison of some CML ring oscillator data shows minimum values of about 40 ps at about 2 mW per gate. With the NTL circuit technique, gate delay as low as 20 ps [1.16] has been obtained. The power per gate at which the minimum of the delay time is found, is still very high in view of complex circuits with complexity higher than about 10 k. This means that one of the main directions for further development has to be the reduction of power dissipation without too much loss of speed. The reduction of the power dissipation, which first implies the reduction of parasitic capacitances, can be done in various ways. The simplest but an expensive way is to scale down the device [1.40,41], meaning that by using advanced lithography and correctly tailored vertical doping profile, the lateral as well as the vertical dimensions are reduced in a coordinated manner. The conditions for optimization are specific for the field of application [1.85]

Another way is to cut down the parasitics by using advanced isolation schemes minimizing the base-collector capacitance, as well as the collector-substrate capacitance. In spite of the effort needed for this approach, several solutions have been presented [1.23-29]. It will become necessary if very dense circuits are to be obtained, as in the case of RAM's.

A third way is to construct completely new transistor schemes like SICOS [1.58-60], the most advanced self-aligned transistor principle presently found.

The most widely used demonstration circuits for digital signal processing besides ring oscillators are static frequency dividers. A static divider consists basically of the well-known series-gated master-slave D-flip-flops with the inverted output fed back to the data input. The number of stages, each consisting of one flip-flop, determines the ratio between input and output frequencies in steps of 2^n (n = 1, 2 3...). The upper frequency limit is determined by the first stage. The main effort for design and optimisation has therefore to be concentrated on this part. As a consequence, the power dissipation of the first stage will be much higher than those of the following stages. This fact should be kept in mind to enable comparisons.

The highest values of working frequency are about 10 GHz with bipolar self-aligning technology [1.86, 87]. For GaAs, values as high as 25.2 GHz have been obtained [1.88], using very sophisticated and expensive laboratory techniques. With a bipolar silicon 2 μm pre-production technology, even 8 GHz have been achieved [1.89]. It should be noted that the power consumption of these circuits is low at reduced frequency demand e.g. 1 GHz dividers with no more than 10 mW power consumption are easily obtained [1.89].

The compatibility for high-speed complex circuits is demonstrated best by complex gate arrays and SRAM's. Complexity of gate arrays ranges up to 18 k [1.43]. The gate delay can be as low as 43 ps for a 2.1K gate array [1.90].

SRAM's between 1 k and 64 k have been achieved with access times ranging from 0.85 ns for 1 k, through 1 ns for 5 k [1.86, 91] to 5 ns for 64 k [1.92].

1.8 Summary and Prospects

In the previous sections, it has been shown that rapid progress is being made in the whole field of bipolar technolgoy, device modelling, and high-speed circuitry. At the moment, no definite limits to this progress are in sight. The potential developments can be roughly classified into several branches:

- With the standard processing techniques presently known, self-aligned device configurations can be scaled down within quite a large range; the device configurations known at present can evolve into new configurations and completely novel device configurations can be found as well, e.g. with complete self-alignment of emitter, base contact, and collector region.

- Employing non-standard processing techniques, e.g. MBE, selective epitaxy or deposition, and the photo-induced deposition techniques, completely new devices may be obtained, even using state-of-the-art device principles. Examples may be devices with box-shaped doping profiles as well as hetero-devices with band gap tailored base and/or emitter.

- Improved or new circuit techniques may demand for improved and/or for new devices. Examples are complementary logic circuit techniques and analogue circuit techniques demanding high-speed pnp transistors.

- Optical and electron-beam circuit testing and measurement techniques may call for integrated optical switches and detectors.

Within the first group, improvements using standard processing will mainly use the scalability of polysilicon emitters. Limits may be given by high-doping effects leading to non-ideality, especially of the input (base current) characteristic.

In the second group, devices based on non-standard processing, important milestones have been already reached. These milestones are primarily hetero-devices of either larger band gap in the emitter or smaller band gap in the base. Serious problems arise mainly from the lattice mismatch and the emitter resistance.

As an example of the third group, the investigations on complementary bipolar logic should be mentioned. With the name CBL (Charge Buffered Logic), a new type of logic element has been proposed [1.93,94] which is characterized by a very low power x delay product at reasonably high speed. With pnp transistors of high transit frequency, $f_T \geq 5$ GHz, power x delay products of <10 fJ and gate delays of < 100 ps should be reached.

These examples illustrate the immense potential for further research and development in the field of silicon bipolar devices and circuits.

References

1.1   S.M. Sze: *Physics of Semiconductor Devices* (Wiley, New York 1981) 2nd ed. Chap.3

1.2   H.-M Rein, R. Ranfft: *Integrierte Bipolarschaltungen*, Halbleiter Elektronik 13 (Springer, Berlin, Heidelberg 1980)

1.3   W. Bräckelmann, A. Glasl, E. Gonauser, W.Wilhelm: A high-speed bipolar masterslice array with 2600 gates. IEEE CCIC Tech. Dig. (1983) pp.27-31

1.4   J. Graul, H. Kaiser, W. Wilhelm, H. Ryssel: Bipolar high speed low-power gates with double implanted transistors, IEEE J. SC-10, 201-204 (1975)

1.5   J. Lohstroh: Devices and Circuits for bipolar (V)LSI Proc. IEEE 69, 812-826 (1981)

1.6   T.H. Ning, R.D. Isaac: Effect of emitter contact on current gain of silicon bipolar devices. IEEE Trans. ED-27, 2051-2055 (1980)

1.7   W.R. Heller, C.G. Hsi, W.F. Mikhail: Wireability-designing wiring space for chips and chip packages. IEEE Des. Test 1, 53-51 (1984)

1.8   W. Klein, E.F. Miersch, R. Remshardt, H. Schettler, U. Schultz, R. Zuhlke: A study on bipolar VLSI gate-arrays assuming four layers of metal. IEEE J. SC-17, 472-480 (1982)

1.9   M. Usami, S. Hososaka, A. Anzai, K. Otsuka, A. Masaki, S. Murata, M. Ura, M. Nakagawa: Status and prospect for bipolar ECL gate array. ICCD Tech. Dig. (1983) pp.272-275

1.10 S.C. Lee, A.S. Bass: A 2500-gate macro cell array with 250 ps gate delay. ISSCC Dig. (1982) pp.178-179

1.11 M. Tatsuki, S. Kato, M. Okabe, H. Yakushiji, M. Teri, T. Noda: A 280 ps/gate ECL 5000-gate array. Proc. IEEE Custom Integrated Circuit Conf. (1985) pp.18-21

1.12 M. Takagi, K. Nakayama, Ch. Terada, H. Kamioka: Improvement of shallow base transistor technology by using a doped polysilicon diffusion source. J. Jpn. Soc. Appl. Phys. (Suppl.) 42, 101-109 (1972)

1.13 J. Graul, A. Glasl, H. Murrmann: High-performance transistors with arsenic-implanted polysil emitters. IEEE J. SC-11, 491-495 (1967)

1.14 T. Sakai, M. Suzuki: Super self-aligned bipolar technology. Symp. VLSI Technology. Dig. (1983) pp.16-19

1.15 S. Konaka, Y. Yamamoto, T. Sakai: A 30 ps Si bipolar IC using super self-aligned process technology. 16th Int'l Conf. on Solid-State Devices and Materials (1984), Extended Abstracts pp.209-212

1.16 S. Konaka, Y. Amemiya, K. Sakuma, T. Sakai: A 20 ps/G Si bipolar IC using advanced SST with collector ion implantation. 19th Int'l Conf. on Solid-State Devices and Materials (1987), Extended Abstracts pp.331-334

1.17 F.S. Becker, H. Oppolzer, I. Weitzel, H. Eichermüller, H. Schaber: Low resistance polycrystalline silicon by boron or arsenic implantation and thermal crystallization of amorphously deposited film. J. Appl. Phys. 56, 1233-1236 (1984)

1.18 T. Tashiro, H. Takemura, T. Kamiya, F. Tokuyoshi, S. Ohi, H. Shiraki, M. Nakamae, T. Nakamura: An 80 ps ECL Circuit with high current density transistor. IEDM Tech. Dig. (1984) pp.686-689

1.19 T.C. Chen, C.T. Chuang, G.P. Li, J.D. Cressler, E.D. Petrillo, S.B. Brodsky, R.N. Schulz, M.R. Polcari, M.B. Ketchen, D.D. Tang: An advanced bipolar transistor with self-aligned ion-implanted base and W/poly emitter. IEEE BCTM Tech. Dig. (1987) pp.31-33

1.20 T.P. Chow, A.J. Steckl: Refractory metal silicides: thin-film properties and processing technology. IEEE Trans. ED-30, 1480-1497 (1983)

1.21 H. Kabza, M. Reisch, V. Probst, W. Böhm, J. Fertsch, H. Schaber, H. Eggers: Double self-aligned bipolar transistors using salicide contacts. Proc. ESSDERC (1987) pp.1047-1050

1.22 S.P. Murarka: Silicides for VLSI Applications (Academic, London 1983) p.30

1.23 H. Yamamoto, O. Mizuno, T. Kubota, M. Nakamae, H. Shiraki, Y. Ikushima: High-speed performance of a basic ECL gate with 1.25 micron design rule. Symp. VLSI Technology, Dig. (1981) pp.38-39

1.24 D.D. Tang, P.M. Solomon, T.H. Ning, R.D. Isaac, R.E. Burger: 1.25 μm deep-groove-isolated self-aligned bipolar circuits. IEEE J. SC-17, 925-931 (1982)

1.25 R.J. Houghton, P.P. Boulay, R.F. Penoyer: A 72 Kb bipolar DRAM. ISSCC Dig. (1982) pp.70-71

1.26 Y. Tamaki, T. Shiba, N. Honma, S. Mizuo, A. Hayasaka: New U-groove isolation technology for high-speed bipolar memory. Symp. VLSI Technology (1983), Dig. pp.24-25

1.27 K. Toyada, M. Tanaka, H. Isogai, C. Ono, Y. Kawabe, H. Goto: A 15 ns 16 Kb ECL RAM with a pnp load cell. ISSCC (1983), Dig. Techn. Papers pp.108-109

1.28 H. Sakai, K. Kikuchi, S. Kameyama, M. Kajiyama, T. Komeda: A trench isolation technology for high-speed and low-power dissipation bipolar LSI's. Symp. VLSI Technology (1987), Dig. Techn. Papers pp.17-18

1.29 G.P. Li, T.H. Ning, C.T. Chuang, M.B. Ketchen, D.D. Tang, J. Mauer: An advanced high-performance trench-isolated self-aligned bipolar technology. IEEE Trans. ED-34, 2246-2254 (1987)

1.30 H.C. de Graaff, G. de Groot: The SIS tunnel emitter: A theory for emitters with thin interface layers. IEEE Trans. ED-26, 1771-1776 (1979)

1.31 T.H. Ning, R.D. Isaac: Effect of emitter contact on current gain of silicon bipolar devices. IEEE Trans. ED-27, 2051-2055 (1980)

1.32 A.W. Wieder: Self-aligned bipolar technology - new chances for very high-speed digital integrated circuits. Siemens Forsch. u. Entwickl. Ber. 13, 246-252 (1984)

1.33 H. Schaber, B. Benna, L. Treitinger, A.W. Wieder: Conduction mechanisms of polysilicon emitters with thin interfacial oxide layers. IEDM Tech. Dig. pp.738-741, 1984

1.34 G.L. Patton, J.C. Bravman, J.D. Plummer: Physics, technology and modelling of polysilicon emitter contacts for VLSI bipolar transistors. IEEE Trans. ED-33, 1754-1768 (1986)

1.35 Z. Yu, B. Ricco, R.W. Dutton: A comprehensive analytical and numerical model of polysilicon emitter contacts in bipolar transistors. IEEE Trans. EC-31, 773-784 (1984)

1.36 A. Neugroschel, M. Arienzo, Y. Komen, R. Isaac: Experimental study of the minority-carrier transport at the polysilicon-monosilicon interface. IEEE Trans. ED-32, 807-816 (1985)

1.37 J.M.C. Stork, M. Arienzo, C.Y. Wong: Correlation between the diffusive and electrical barrier properties of the interface in polysilicon contacted n^+-p junctions. IEEE Trans. ED-32, 1776-1770 (1985)

1.38 B. Benna, T.F. Meister, H. Schaber: The role of the interfacial layer in bipolar polysilicon emitter transistors. Solid-State Electron. 30, 1153-1158 (1987)

1.39 J.M.C. Stork, J.D. Cressler: Performance degradation due to emitter resistance in polysilicon emitter bipolar transistors. Symp VLSI Technology (1986), Dig. Techn. Papers pp.47-48

1.40 D.D. Tang, P.M. Solomon: Bipolar transistor design for optimized power-delay logic circuits. IEEE J. SC-14, 679-684 (1979)

1.41 P.M. Solomon, D.D. Tang: Bipolar circuit scaling. ISSCC (1979), Dig. Tech. Papers pp.86-87

1.42 C. Volz, L. Bloßfeld: Collector implanted technology for Si bipolar devices. Proc. ESSDERC (1986) pp.136-137

1.43 T. Nishimura, H. Sato, M. Tatsuki, T. Hirao, Y. Kuramitsu: A bipolar 18K-gate variable size cell master-slice. IEEE J. SC-21, 727-732 (1986)

1.44 J. Fertsch, H. Klose, W. Böhm: Base resistance calculation for high speed bipolar transistors. Proc. ESSDERC (1986) pp.121-122

1.45 J. Fertsch, H. Voit, K. Klose, W. Böhm: Measurements and calculation of base-resistance components of modern high-speed bipolar transistors. Proc. ESSDERC (1987) pp.969-972

1.46 D.D. Tang, G.P. Li, C.T. Chuang, T.H. Ning: On the impurity profiles of down-scaled bipolar transistors. IEDM (1986), Tech. Dig. pp.412-415

1.47 C.T. Chuang: Transient imposed scaling considerations in advanced narrow-emitter bipolar transistors. IEDM (1987), Tech. Dig. pp.178-181

1.48 A. Gover, A. Gaash: Experimental model aid for planar design of transistor characteristics in integrated circuits. Solid-State Electron. 19, 125-127 (1976)

1.49 H.-M. Rein: A simple method for separation of the internal and external (peripheral) currents of bipolar transistors. Solid-State Electron. **27**, 625-631 (1984)

1.50 M. Miura-Mattausch: Current gain dependence on the emitter size of polysilicon-emitter bipolar transistor. Proc. ESSDERC (1987) pp.909-912

1.51 C.A. Grimbergen: The influence of geometry on the interpretation of the current in epitaxial diodes. Solid-State Electron. **19**, 1033-1037 (1976)

1.52 P.-T.Chen, K. Misiakos, A. Neugroschel, F.A. Lindholm: Analytical solution for two-dimensional current injection from shallow p-n junctions. IEEE Trans. ED-32, 2292-2296 (1985)

1.53 Y. Tamaki, F. Murai, K. Sagara, A. Anzai: A 100 nm emitter transistor fabricated with direct EB writing for high-speed bipolar LSIs. Symp. VLSI Technology (1987), Dig. Tech. Papers pp.31-32

1.54 R.S. Muller, T.I. Kamins: *Device Electronics for Integrated Circuits* (Wiley, New York 1977) Chap.5

1.55 C.T. Kirk: A theory of transistor cutoff frequency fall-off at high current density. IEEE Trans. ED-9, 164-174 (1962)

1.56 J.M.C. Stork, R.D. Isaac: Tunnelling in base-emitter junctions. IEEE Trans. ED-30, 1527-1534 (1983)

1.57 J. del Alamo, R.M. Swanson: Forward-bias tunnelling current limits in scaled bipolar devices. 18th Int'l Conf. on Solid-State Devices and Materials (1986), Extended Abstracts pp.283-286

1.58 T. Nakamura, T. Miyazaki, S. Takahashi, T. Kure, T. Okabe, M. Nagata: Self-aligned transistor with sidewall base electrode. IEEE Trans. ED-29, 596-600 (1982)

1.59 K. Nakazato, T. Nakamura, M. Kato: A 3 GHz lateral pnp transistor. IEDM (1986), Techn. Dig. pp.416-419

1.60 Y. Tamaki, T. Shiba, K. Ikeda, T. Nakamura, N. Natsuaki, S. Ohyo, T. Hayashida: New self-aligned bipolar device process technology for sub-50 ps ECL circuits. IEEE BCTM (1987), Techn. Dig. pp.22-23

1.61 H. Kroemer: Heterostructure bipolar transistors and integrated circuits. Proc. IEEE **70**, 13-25 (1982)

1.62 B.L. Sharma, R.K. Purohit: *Semiconductor Heterojunctions* (Pergamon, Elmsford 1974)

1.63 A.Y. Cho, J.R. Arthur: Molecular Beam Epitaxy. Prog. Solid State Chem. **10**, 157-191 (1975)

1.64 K. Ploog: Molecular beam epitaxy of III-V compounds, in *Crystals: Growth, Properties and Applications* **3**, 73-162 (Springer, Berlin, Heidelberg 1980)

1.65 R.D. Dupuis, L.A. Moudy, P.D. Dapkus: Preparation and Properties of $Ga_{1-x}Al_x$As-GaAs Heterojunctions grown by metal-organic chemical vapor deposition. Gallium Arsenide and Related Compounds (St. Louis), Inst. Phys. Conf. Ser. **45**, 1-9 (1979)

1.66 R. Mertens, J. Nijs, J. Symons, K. Baert, M. Ghannam: Comparison of advanced emitters for high speed silicon bipolar transistors. IEEE BCTM (1987), Tech. Dig. pp.54-56

1.67 M. Ghannam, J. Nijs, R. Mertens, R. De Keersmaecker: A silicon bipolar transistor with a hydrogenated amorphous emitter. IEDM (1984), Tech. Dig. pp.746-748

1.68 H. Matsunami, S. Nishino, H. Ono: Heteroepitaxial growth of cubic silicon carbide of foreign substrates. IEEE Trans. ED-28, 1235-1236 (1981)

1.69 T. Sugii, T. Ito, Y. Furumura, M. Doki, F. Mieno, M. Meada: Epitaxial SiC emitter for high speed bipolar VLSIs. Symp. VLSI Technology (1986), Dig. Techn. Papers pp.45-46

1.70 T. Sugii, T. Ito, Y. Furumura, M. Doki, F. Mieno, M. Maeda: Si heterojunction bipolar transistors with single-crystalline β-SiC emitters. J. Electrochem. Soc. **134**, 2545-2549 (1987)

1.71 T. Sugii, T. Ito, Y. Furumura, M. Doki, F. Mieno, M. Maeda: β-SiC/Si heterojunction bipolar transistors with high current gain. IEEE EDL-9, 87-89 (1988)

1.72 K. Sasaki, S. Furukawa: A micro-crystal emitter heterojunction bipolar transistor. 19th Int'l Conf. on Solid-State Devices and Materials (1987), Extended Abstracts pp.335-338

1.73 M. Tabe, M. Takahashi, Y. Sakakibara: Oxygen-doped Si epitaxial films for Si hetero-bipolar transistors. 18th Int'l Conf. on Solid-State Devices and Materials (1986), Extended Abstracts pp.37-40

1.74 Second Int'l Symp. on Si Molecular Beam Epitaxy, Honolulu (1987)

1.75 J.C. Bean, L.C. Feldman, A.T. Fiory, S. Nakahara, I.K. Robinson: Ge_xSi_{1-x}/Si strained-layer superlattice grown by molecular beam epitaxy. J. Vac. Sci. Technol. A 2, 436-440 (1984)

1.76 D.V. Lang, R. People, J.C. Bean, A.M. Sergent: Measurement of the bandgap of Ge_xSi_{1-x}/Si strained-layer heterostructures. Appl. Phys. Lett. **47**, 1333-1335 (1985)

1.77 S.S. Iyer, R.A. Metzger, F.G. Allen: Sharp doping profiles with high and low doping levels in silicon grown by molecular beam epitaxy. J. Appl. Phys. **52**, 5608-5613 (1981)

1.78 C. Smith, A.D. Welbourn: Prospects for a hetero-structure bipolar transistor using a silicon-germanium alloy. IEEE BCTM (1987), Tech. Dig. pp.57-60

1.79 H.C. de Graaff: Compact bipolar transistor modelling, in *Process and Device Modelling*, ed. by W.L. Engl (North-Holland, Amsterdam 1986) pp.413-432

1.80 W.L. Engl, O. Manck, A.W. Wieder: Modelling of bipolar devices, in *Process and Device Modelling for Integrated Circuit Design*, ed. by F. van de Wiele, W.L. Engl, P.G. Jespers (Noordhoff, Leyden 1977) pp.377-418

1.81 H.C. de Graaff, G.A.M. Hurkx: Physical modelling problems of ultrafast silicon bipolar transistors. Proc. ESSDERC (1987) pp.503-506

1.82 R.W. Knepper: Problems in high performance bipolar device modelling. IEEE BCTM (1987), Tech. Dig. pp.1-4

1.83 R.W. Knepper, S.P. Gaur, F.-Y. Chang, G.R. Srinivasan: Advanced bipolar transistor modelling: Process and device simulation tools for today's technology. IBM J. Res. Develop. **29**, 218-228 (1985)

1.84 H.R. Claessen, J.A.M. Geelen, H.C. de Graaff: The influence of emitter sidewall injection on transistor noise figure. ESSDERC (1987) pp.897-900

1.85 W. Wilhelm: Characterization and optimisation of bipolar technologies by means of high-speed circuit design. Proc. ESSDERC (1987) pp.499-501

1.86 H. Sakai, S. Konaka, Y. Yamamoto, M. Suzuki: Prospects of SST technology for high speed LSI. IEDM (1985), Tech. Dig. pp.18-21

1.87 K. Washio, T. Nakamura, K. Nakazato, T. Hayashida: A 48 ps ECL in a self-aligned bipolar technology. ISSCC (1987), Dig. Tech. Papers, pp.58-59

1.88 J.F. Jensen, V.K. Mishra, A.S. Brown, R.S. Beaubien, M.A. Thompson, R. Jelloian: 25 GHz static frequency dividers in AlInAs-GaInAs HEMT technology. ISSCC (1988), Dig. Techn. Papers, pp.268-269

1.89 P. Weger, H.-M. Rein: Speed-power relation of modern bipolar technology. Proc. ESSDERC (1987), pp.1051-1054

1.90 M. Suzuki, M. Hirata, S. Konaka: 43 ps/5,2 GHz bipolar macrocell array LSI's. ISSCC (1988), Dig. Techn. Papers, pp.70-71

1.91 C. Chuang, D.D. Tang, G.P. Li, E. Hackbarth, R.R. Boedecker: A 1.0-ns 5-k bit ECL RAM. IEEE J. SC-21, 670-674 (1986)

1.92 T. Awaya, K. Toyada, O. Nomura, Y. Nakaya, K. Tanaka, H. Sugawara: A 5 ns access time 64 kb ECL RAM. ISSCC (1987), Dig. Tech. Papers, pp.130-131

1.93 S.K. Wiedmann, H.H. Berger: Bipolar complementary logic (CTL). Proc. 1st Europ. Solid-State Circuit Conf. (1975) pp.36-39

1.94 S.K. Wiedmann: Charge buffer logic (CBL) - A new complementary bipolar circuit concept. Symp. VLSI Technology (1985), Dig. Techn. Papers, pp.38-39

2. Self-Aligning Technology for Sub-100nm Deep Base Junction Transistors

Masahiko Nakamae
VLSI Development Division, NEC Corporation
1120, Shimokuzawa, Sagamihara
Kanagawa 229, Japan

This chapter describes the newly developed technology which will break through the limitations for further scaling down of the base junction depth of a self-aligned bipolar transistor. The new technology, BSA (BSG Self-Aligned) technology, features the use of BSG film as a sidewall spacer between the emitter and base electrodes as well as the diffusion source for the intrinsic base and also for the p^+-link region between the intrinsic and extrinsic base. BSA technology has been successfully combined with the RTA (Rapid Thermal Annealing) technique to fabricate sub-100nm base self-aligned bipolar transistors. The typical BSA transistor has $h_{FE} = 70$, $BV_{CEO} = 7$ V and $BV_{EBO} = 3$ V. BSA technology is likely to prove extremely useful in future bipolar VLSIs.

2.1 Background

It is well known that the recent progress in high-speed performance of Si bipolar devices has mainly been achieved by modern self-aligning technologies [2.1, 2].

Figure 2.1 shows a schematic cross-sectional view of a typical modern self-aligned transistor. This transistor has several outstanding features which make it possible to realize very high speed performance. First, a very narrow emitter width (shown as "a") of about 0.5 to 0.3 μm is easily obtained even in 1 μm design rule. This leads to the considerable reduction of the intrinsic base resistance and emitter-base junction capacitance. Emitter and base poly-Si electrodes are separated by the sidewall spacer (shown as "b"). The separation is typically 0.2 to 0.3 μm, resulting in a very low base resistance.

In the case of the most advanced self-aligned transistor, the ultimately small base area ("c") of less than 2 μm, can be realized. Therefore, the drastic reduction of collector-base junction capacitance is achieved. Furthermore, the poly-Si emitter ("d") provides sufficient current gain in vertically scaled-down transistors without suffering from collector-emitter punchthrough. To achieve higher performance,

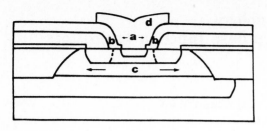

a : 0.5 to 0.3um using 1um design rule

b : side wall spacer

c : ultimately small base area

d : poly Si emitter

Fig.2.1. Schematic cross-sectional view of typical modern self-aligned bipolar transistor

increase of the cutoff frequency is the major concern according to the device parameter analysis. Therfore, an ultra shallow base junction depth, such as sub-100nm, is required to obtain a cutoff frequency above 25 GHz at V_{CE}=1 V. However, it is quite difficult to utilize the present technologies for further scaling down, because of several limitations in process and device issues.

2.2 Limitations of the Modern Self-Aligned Transistors

2.2.1 Process Issues

Usually the low-energy ion implantation method [2.3] has been utilized to form the shallow base region. But the secondary channelling effect becomes a severe problem if one wishes to realize a sub-100nm base junction depth even at a very low energy.

During the boron implantation into a silicon crystal for base formation, the wafer is usually tilted about 7°, sometimes with a rotation, to prevent the boron ions from traveling along the crystalline channels. However, some portion of the boron ions may still be scattered into the channels (secondary channeling). Thus a channeling tail should always exit in the boron distribution profile, as shown in Fig.2.2. In addition, the crystal angle for the channeling increases as the ion energy decreases. Then, a large portion of the boron ions may enter the channels for low-energy ion implantation. As a result, this tailing profile limits the realization of sub-100nm base transistors.

Another severe problem is the formation of a very shallow and highly doped base. Figure 2.3 depicts the residual lattice defect density versus implanted dose at a given acceleration energy and post-annealing conditions. The defect density is almost zero until the dose reaches some critical value. But, at this critical dose, the defect density starts to increase with increasing dose, and reaches a maximum value before decreasing again to zero.

secondary channeling effect

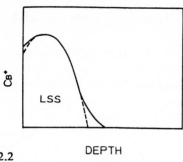

lattice defect

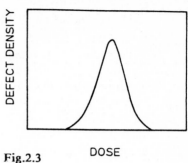

Fig.2.2

Fig.2.3

Fig.2.2. Schematic boron profiles of the LSS theory and SIMS measurements

Fig.2.3. Schematic drawing of residual lattice defect density as a function of implanted dose

This defect annihilation is due to the recrystallization of the amorphous layer formed by the high dose implantation. Unfortunately, a boron dose of this critical value or more must be used in a very shallow base to avoid the emitter-collector punchthrough.

In general, high-temperature annealing is quite effective to annihilate the implantation damage. However, it cannot be applied for the shallow base formation. The residual damage, therefore, should exist in the base region and cause the degradation of the transistor characteristics.

2.2.2 Device Issues

In a self-aligned, very shallow base transistor, the overlap of the internal and external base region is the main design consideration to avoid the emitter perimeter punchthrough and to achieve very-high-speed performance.

In present technology, the overlap is controlled by the lateral diffusion of the external base region. However, this method causes several problems:

a) The emitter-base tunnelling current is considerably increased at the emitter perimeter in order to ensure the overlap.

b) In the above-mentioned case, the vertical junction depth of the external base is increased and, therefore, the collector-base junction capacitance is increased. As the area ratio of the extrinsic base is more than 3 in a modern self-aligned transistor, this is quite harmful to achieve high-speed performance.

c) Degradation of h_{FE} and f_T under high injection conditions occur in the emitter perimeter region.

It is necessary to increase the lateral diffusion length for shallower base transistors. These problems are becoming more severe as the base junction depth are scaled. If the sidewall width of the modern self-aligned transistor can be scaled, we need not worry about the problems. But the sidewall width has a lower limit determined by the isolation voltage of the sidewall insulator at the emmitter-base capacitance between the polysilicon electrodes. Another problem should be encountered when the emitter width is further scaled. That is, in the conventional boron ion implantation technology, shadowing effect should occur at one side of the emitter window having a high aspect ratio [2.4]. This shadowing effect causes insufficient overlap of the external and the internal base region, resulting in the emitter perimeter punchthrough.

These incompatible design requirements strongly limit the further progress of high speed bipolar transistors.

The situation is the same as the case of shallow source-drain formation in submicron p-channel MOSFETs. To overcome these obstacles, various methods have been investigated in recent years. In the following, several methods are reviewed briefly.

2.2.3 Preamorphization by Si⁺ Implantation

In this technique, the silicon surface is preamorphized to a depth of about 0.3 μm, considerably beyond the intended junction depth of about 0.1 μm, for one case [2.5]. Then, BF_2^+ molecules are implanted to give a low equivalent boron energy. For example, 50 keV BF_2^+ gives about 11 keV B^+ because the mass ratio of boron and the BF_2^+ molecule is 0.22. The typical boron and summed disorder depth profiles are shown in Fig.2.4.

It is noticed that the depth of the amorphous/crystalline interface, in this case about 0.3 μm, should be located outside the space-charge region.. Another case of the preamorphization study is shown in Fig.2.5. In this case [2.6], when the amorphous/crystalline interface located near the base junction, the reverse biased p^+n junction leakage current increases abruptly at a critical energy of the Si implantation, related to the boron profile.

In order to realize high-performance LSIs using this technology, it is most important to eliminate the defects existing near the interface.

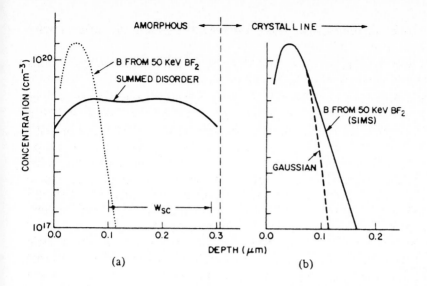

(a)   (b)

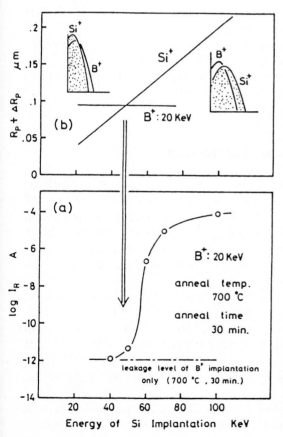

Δ

Fig.2.4a,b. The B depth profile, with the lack of a channeling tail, is shown for 50keV BF_2^+ implanted into amorphous silicon (**a**). In addition, the summed relative disorder produced by multiple Si^+ implants, and the space-charge width W_{SC} for 2V reverse bias junction in annealed silicon are shown in (**a**). (**b**) The typical B depth profile with channel tail is shown for BF_2^+ implanted into crystalline silicon [2.5]

Fig.2.5a,b. Effect of Si^+ implantation energy on (**a**) implanted silicon distribution relative to 20kV implanted boron distribution with the same dose, $5 \cdot 10^{15}$ cm^{-2}; (**b**) reverse biased p$^+$n junction leakage current. (5V; $200 \times 480 \mu m^2$) [2.6]

33

2.2.4 Low–Temperature Photo–Epitaxy Process

The epitaxial-growth temperature has considerably been reduced, to as low as $540°C$ using a Si_2H_2 gas source and ultra-violet light irradiation which enhances the dissociation of Si_2H_6 gas and the surface migration of Si atoms [2.7]. In situ boron doping is carried out for a 0.1 μm epitaxial base thickness and a $1 \cdot 10^{19}$ cm^{-3} peak concentration at a growth temperature as low as $650°C$.

This low-temperature in-situ doping process can minimize the redistribution of dopant.

The intrinsic base region and the polysilicon base electrode are fabricated at the same time using this process, the base-collector capacitance can be minimized in a self-aligned manner.

Figure 2.6 shows the main process steps to fabricating the epitaxially-grown base transistor (EBT) using the low-temperature photo-epitaxy process.

Figure 2.7 exhibits the SIMS depth profiles [2.8] of an intrinsic transistor region.

(a) Field oxidation

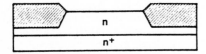

(b) Photo–epitaxy (Boron doped)

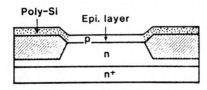

(c) Poly–Si CVD (Arsenic doped)

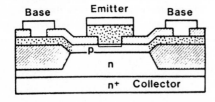

Fig.2.6. The epitaxially grown base transistor (EBT) fabrication steps [2.7]

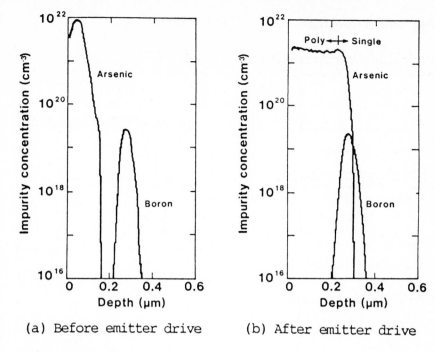

(a) Before emitter drive   (b) After emitter drive

Fig.2.7. SIMS depth profiles of an intrinsic transistor region of the EBT [2.7]

2.2.5 Gallium Diffusion Process

A shallow and heavily doped junction has been formed from a Ga-implanted SiO_2 films [2.9].

In this process, Ga is introduced into the Si crystal by a thermal diffusion process, instead of ion implantation, resulting in no channeling tail and crystal damage.

Ga diffuses so fast into SiO_2 film that Ga immediately reaches the SiO_2/Si interface and diffuses into Si very slowly. At the interface Ga segregates in the Si side. These characteristics of Ga are favorable for a shallow and heavily doped base formation.

The base doping concentration is controlled by the Ga implantation conditions into SiO_2 film.

Figure 2.8 sketches the main fabrication steps of the Ga base transistor.

Figure 2.9 illustrates the example of Ga profile in the Si_3N_4/SiO_2/Si structure.

35

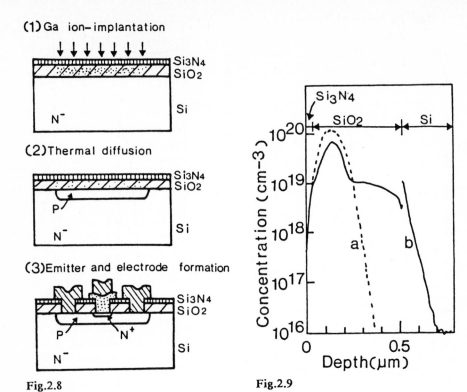

(1) Ga ion-implantation

Si3N4
SiO2

Si

N⁻

(2) Thermal diffusion

Si3N4
SiO2

P

Si

N⁻

(3) Emitter and electrode formation

Si3N4
SiO2

P   N⁺

Si

N⁻

Fig.2.8

Fig.2.9

Fig.2.8. Fabrication steps of gallium base transistor [2.9]

Fig.2.9. Typical example of gallium diffusion in $Si_3N_4/SiO_2/Si$ structure. Gallium dose: $1\cdot10^{15}$ cm^{-2}, 160 keV; Si_3N_4: 300 Å, SiO_2: 5000 Å a: before heat treatment, b: after 900°C-30min heat treatment [2.9]

2.2.6 Laser Doping

The gas immersion laser doping (GILD) process involves placing the wafer into the pyrolytic B_2H_6 gas cell and irradiating with a XeCl excimer laser [2.10].

The dopant gas B_2H_6 is absorbed on the Si surface and then driven in during a melting/regrowth process initiated upon exposure to the short laser pulse. In this process, the junction depth and the boron concentration are controlled by the laser energy and the number of laser pulses, respectively.

Figure 2.10 displays the cross section to explain the doping mechanism of the GILD process. Here, Al is used for the reflecting film.

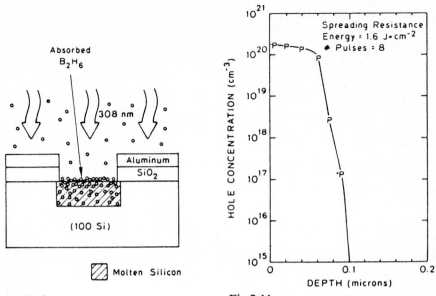

Fig.2.10   **Fig.2.11**

Fig.2.10. Cross section of a masked (100) wafer showing the adsorption of the dopant species (in this case B_2H_6) onto the clean silicon surface. The dopant is incorporated into the very shallow region upon exposure to the excimer laser pulse

Fig.2.11. Spreading resistance profile of active hole concentration showing a p^+-laser-doped shallow junction [2.10]

Figure 2.11 exhibits the whole concentration profile obtained under a typical laser irradiation condition.

2.3 BSA Technology

In the following, a novel self-aligning technology, BSA (BSG Self-Aligned) is demonstrated. Figure 2.12 shows the schematic cross section of the BSA transistor. The BSA technology is characterized by the following:

a) A BSG film is employed as a sidewall spacer between the emitter and base electrodes.

b) The BSG film is also used as the diffusion source to form the intrinsic base and the p^+-link region which is the overlap region of the intrinsic and extrinsic base.

37

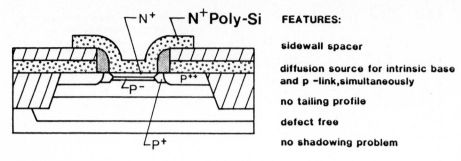

FEATURES:

sidewall spacer

diffusion source for intrinsic base
and p –link, simultaneously

no tailing profile

defect free

no shadowing problem

Fig.2.12. Schematic cross section of a BSA transistor

Moreover, the boron profile shows no tailing characteristics, and a defect-free base region can be formed even at high boron concentrations. In addition, the shadowing problem does not occur even in an ultimately narrow emitter window.

2.3.1 Process Flow

The basic fabrication process steps of the BSA transistor are explained schematically in Fig.2.13 and described below:

a) The emitter window is opened in the SiO_2/poly-Si stacked layer deposited on the oxide-isolated epitaxial layer. After the drive-in process, a shallow extrinsic base region is formed in the epitaxial layer. Then, BSG film is deposited on the surface. The thickness of the BSG film is 0.2 to 0.3 μm.

b) RTA under typical conditions (1000°C for 10s) is applied to form the sub-100nm base. Here, the overlap between the intrinsic and extrinsic base is perfect.

c) The BSG film is etched using the reactive ion etching technique to form the emitter contact and the sidewall spacer.

d) The emitter poly-Si is deposited, and arsenic ions are implanted into the poly-Si. RTA is then applied again. An extremely shallow emitter is formed at this step. It should be noted that the intrinsic base and the overlap region just under the BSG sidewall spacer receive RTA once more, but that the doping concentration under the sidewall spacer is relatively high. This is because the BSG sidewall spacer is again used as a diffusion source.

Therefore, the optimization of the boron profiles in both regions may be successfully performed by controlling the boron concentration in the BSG film, and by the first and second RTA conditions.

(a) BSG Deposition

Fig.2.13. Basic fabrication steps of the BSA transistor

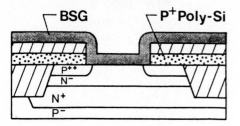

(b) Base Anneal by RTA

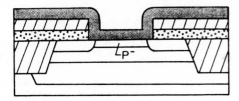

(c) BSG RIE

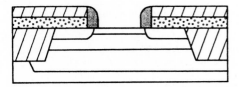

(d) Emitter Anneal by RTA

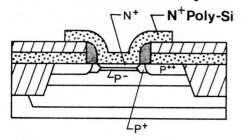

2.3.2 Experimental Results

In this experiment, a BSG film was deposited by the APCVD technique using a B_2H_6 and SiH_4 gas system. Figure 2.14 shows the sheet resistance of the diffused layer versus the boron concentration in the BSG film for the RTA conditions of 1000°C for 10 s. It is necessary to achieve a sheet resistance of about 5 kΩ/square for a very shallow intrinsic base before emitter drive-in.

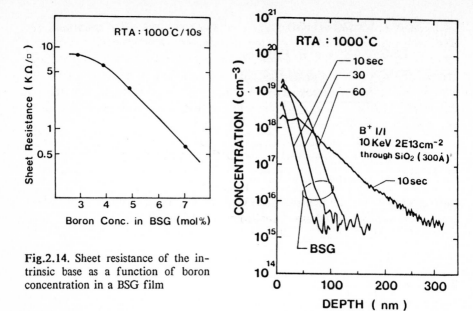

Fig.2.14. Sheet resistance of the intrinsic base as a function of boron concentration in a BSG film

Fig.2.15. Boron profiles obtained by BSA technology and low-energy ion implantation technique

Table 2.1. BSA transistor parameters

Emitter area (mask size)	A_e	1.0×8.0 μm^2
Emitter-base capacitance	C_{je}	24 fF
Collector-base capacitance	C_{jc}	15 fF
Collector-substrate capacitance	C_{js}	30 fF
Base resistance	R_b	50 Ω
Collector resistance	R_c	35 Ω
Current gain	h_{FE}	70
C-E breakdown voltage	BV_{CEO}	7.0 V
E-B breakdown voltage	BV_{EBO}	3.0 V
C-B breakdown voltage	BV_{CBO}	15.0 V

In this case, the optimum boron concentration in the BSG film was determined to be 4.5 mol.%. The boron concentration in the BSG film was controlled by changing the flow ratio of the B_2H_6 and SiH_4 gases.

Figure 2.15 shows ultimately shallow boron profiles obtained by BSA technology and also by a low-energy ion implantation technique for comparison. In this experiment, the epi-doping concentration was about $1.5 \cdot 10^{16}$ cm^{-3}, and a very shallow base junction, 50 to 110 nm, was obtained by changing the RTA time, as shown in the figure. In this SIMS analysis, the surface concentration was not yet saturated after 10 seconds, but reached saturation in 30 seconds.

On the other hand, in a low-energy ion implantation methods, considerable tailing was observed, and the junction depth was as deep as 200 nm.

Table 2.1 shows typical BSA transistor parameters for a 1.0 x 8.0 μm^2 emitter size (mask size). Fairly good dc characteristics were obtained with sub-100nm base transistors and f_T = 22 GHz was obtained by 3D-simulation at a V_{CE} of 1 V.

2.4 Summary

To achieve very-high-speed performance, severe problems for further scaling down the base depth must be solved. These involve both process and device aspects.

In this chapter, a novel self-aligned technology, BSA technology, is demonstrated. The BSA technology features the use of a BSG film as a side-wall spacer and also as a diffusion source to form both the intrinsic base and the p$^+$-link regions at once. The fabricated BSA transistor shows fairly good dc characteristics and a very high f_T value of 22 GHz is estimated.

BSA technology is likely to prove extremely useful for future bipolar devices.

References

2.1   T. Sakai, S. Konaka, Y. Kobayashi, M. Suzuki, Y. Kawai: Gigabit logic bipolar technology: Advanced super self-aligned process technology. Electron Lett. **19**, 283-284 (1983)
2.2   K. Washio, T. Nakamura, K. Nakazato, T. Hayashida: A 48ps ECL in a self-aligned bipolar technology. ISSCC (1984), Dig. Techn. Papers, pp.58-59
2.3   H. Ryssel, H. Glawischnig (eds.): *Ion Implanation Techniques*, Springer Ser. Electrophys., Vol.10 (Springer, Berlin, Heidelberg 1982)
2.4   C.T. Chuang, G.P. Li, T.H. Ning: On the sidewall shadowing effect of self-aligned transistors. Proc. IEEE Bipolar Circuits and Technology Meeting (1987) pp. 24-25
2.5   T.E. Seidel: Rapid thermal annealing of BF$_2$$^+$ implanted preamorphized silicon. IEEE EDL-4, 353-355 (1983)

2.6   K. Yamada, M. Kashiwagi, K. Taniguchi: Formation of shallow p^+n junction by low temperature annealing. Proc. 14th Conf. on Solid State Devices (Tokyo 1982) pp.157-160

2.7   T. Sugii, T. Yamazaki, T. Fukano, T. Ito: Thin-base bipolar technology by low-temperature photo epitaxy. Symp. VLSI Technology (1987) pp.35-36

2.8   H. Oechsner (ed.): *Thin Film and Depth Profile Analysis*, Topics Curr. Phys., Vol.37 (Springer, Berlin, Heidelberg 1984)

2.9   M. Ugajin, S. Konaka, Y. Amemya: A nanometer-base silicon bipolar transistor using an ultra-shallow gallium diffusion process. 19th Conf. on Solid State Devices and Materials (Tokyo 1987) Extended Abstracts, pp.339-342

2.10 P.G. Carey, T.W. Sigmon, R.L. Press, T.S. Fahlen: Ultra-shallow high-concentration boron profiles for CMOS processing. IEEE EDL-6, 291-293 (1985)

3. Vertical Scaling Considerations for Polysilicon-Emitter Bipolar Transistors

T.F. Meister, H. Schaber, K. Ehinger, J. Bieger, B. Benna, and I. Maier

Siemens AG, Central Research and Development
Microelectronics, Otto-Hahn-Ring 6
D-8000 München, Fed. Rep. Germany

Until the early 1980s, the industrial standard of high-speed bipolar processes was characterized by implanted base and arsenic implanted, metal-contacted emitter devices. One of the essential achievements leading to the "renaissance" of bipolar technology [3.1] that has occurred since then is the use of polycrystalline silicon (poly-Si) as diffusion source and contact material for the emitter. Besides facilitating the second important achievement, self-alignment between emitter and base contact, the poly-Si emitter has the following major advantages:

i) Without compromising current gain, extremely shallow emitter junction depths can be formed, leading to a strong reduction in emitter charge storage.

ii) Due to the shallow emitter junction, extremely narrow base regions can be realized with sufficient process control. This, of course, serves to reduce base transit time.

iii) The polysilicon layer interposed between the active emitter region and the metal contact enhances device yield considerably due to its gettering capability and its blocking action against metal sintering and spiking.

After describing the present status and understanding of poly-Si emitters (Sect.3.1) and the corresponding methods to achieve narrow base regions (Sect.3.2), the impact of these achievements on intrinsic device speed will be discussed (Sect.3.3). Based on experimental data and device simulations, the potential for further speed improvement of poly-Si emitter bipolar transistors will be assessed.

3.1 Polysilicon Emitters

Polycrystalline silicon is now widely used in bipolar technology as contacts for the emitter and base of the transistor. In comparison to conventional metal-contacted emitters, improvements in current gain between a factor of 2 [3.2] and 30 [3.3, 4] have been reported for poly-silicon emitter transistors, depending on processing conditions such as the treatment of the interface between the poly- and monocrystalline part of the emitter.

Controversy exists regarding the mechanisms causing a reduction of the base current in polysilicon emitter transistors. *Ning* and *Isaac* [3.2] explained this improvement in terms of a low minority-carrier mobility in the polysilicon layer while *De Graaff* and *De Groot* [3.3] suggested that an oxide layer is formed at the interface acting as a tunnelling barrier to hole transport. Additionally *Chung* and *Yang* [3.5] claimed that a hole barrier is formed due to arsenic segregation at the poly/mono-Si interface and that hole transport across the interface is due to thermionic emission. There has been no general agreement about which class of model best describes the polysilicon emitter transistor. Experimental work [3.6, 7] however has shown that current gain can be improved significantly by the presence of a deliberately grown interfacial oxide layer, suggesting that the tunnel barrier model is valid at least in these transistors.

Because of a significantly increased emitter resistance [3.7, 8] this type of polysilicon transistor is not useful for VLSI-applications. Therefore to minimize oxide contamination of the interface between the single- and poly-crystalline part of the emitter usually a HF-dip etch is performed immediately prior to inserting the wafer into the polysilicon-deposition system. Despite this treatment a thin "native" oxide layer with a thickness of about 10 Å exists at the interface.

3.1.1 Emitter Modelling

At low and intermediate current levels the polysilicon contact does not affect the collector current but reduces the base current which is, however, also influenced by transport effects in the monocrystalline part of the emitter. To extract quantitative information on the polysili-con contact alone, we introduce the effective surface recombination velocity S, defined by [3.9, 10]

$$j_p(x_0) = qS \cdot \Delta p(x_0) \tag{3.1}$$

where x_0 is the position of the interface. In (3.1) $\Delta p(x_0)$ is the excess hole density at $x = x_0$ and $j_p(x_0)$ the hole current density arriving at the interface. Generally S is a lumped parameter containing both the properties of the interface as well as those of the polysilicon layer. We analyzed measured base currents by solving - for different S values as boundary conditions - the complete set of transport equations in the monocrystalline part of the emitter. To determine surface recombination velocity this parameter is varied until measured and calculated base currents coincide [3.11]. This has been done using the device analysator programme MEDUSA [3.12]. In our simulations we used Slotboom's formula for bandgap narrowing [3.13] and the Auger coefficient, as given by *Dziewior* and *Schmid* [3.14]. Additionally we assumed the mobilities of minority and majority carriers to be the same.

As an example Fig.3.1 shows calculated and measured base current densities J_{BO} as a function of the emitter junction depth X_{jE}. For very deep emitters, base saturation currents are independent of X_{jE}. This results from the fact that all holes have recombined in the single-crystalline part of the emitters before reaching the interface. In the case of a metal-contacted emitter ($S = 1 \cdot 10^7$ cm/s) the base current increases with decreasing emitter junction depth indicating that the emitter becomes more and more transparent to hole transport. Therefore reducing X_{jE} in conventional metal-contacted devices leads to unacceptably high base currents. In contrast the "ideal" polysilicon contact with $S = 0$ imposes that no hole crosses the interface and recombination only occurs in the single-crystalline part of the emitter, resulting in low base currents at small values of X_{jE}. The intermediate curve with $S = 6 \cdot 10^4$ cm/s is obtained for a polysilicon contact as cur-

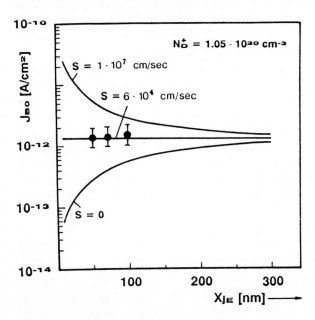

Fig.3.1. The base saturation current density J_{BO} as a function of emitter junction depth X_{jE} for different interface conditions

rently used in VLSI applications. In this case "Rapid Optical Annealing" (ROA) at T = 1050°-1100°C is applied for the emitter drive-in. The base current of these devices is only weakly dependent on X_{jE}. Therefore in polysilicon emitter transistors, emitter junction depth can be scaled down vertically without loss in current gain.

3.1.2 Base Currents for Various Annealing Cycles

The surface recombination velocity, and thus the base saturation current density J_{BO}, depends on interface treatments as well as on annealing procedures. In order to investigate the dependence of the surface recombination velocity and of the emitter resistance on annealing cycles, we report results of devices annealed at different temperatures and times, resulting in emitter junction depths ranging from 40 to 150 nm. The emitters are either formed by "rapid optical annealing" at high temperatures T = 1050°-1100°C and for short times, or by conventional furnace anneals at lower temperatures and longer times. All samples were given the same HF-dip etch prior to the polysilicon dep-

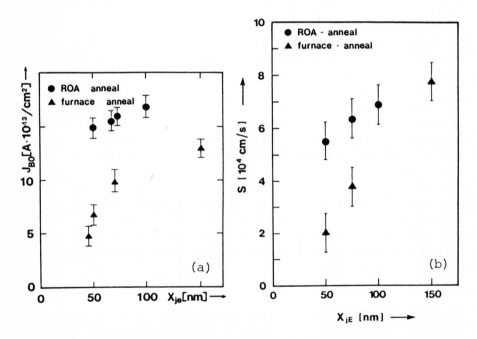

Fig.3.2a,b. (a) Base saturation current densities J_{BO}, and (b) effective surface recombination velocities for a range of emitter drive in temperatures

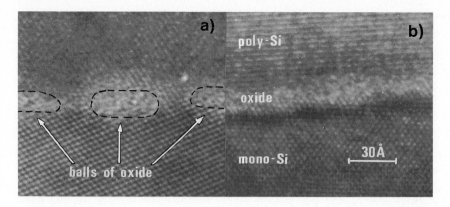

Fig.3.3a,b. XTEM of the poly/mono-Si interface: (a) 1100°C (5s) (b) 900°C/10min emitter drive in

osition. The electrically active dopant concentration in the single-crystalline part of the emitter was estimated by sheet resistance and SIMS measurements [3.11], using the mobility formular of *Arora* et al. [3.15]. The base currents of devices with an active emitter area of 16 x 16 μm^2 were measured at T = 300 K, and surface recombination velocities are extracted in the manner as described before. The results are shown in Figs.3.2a,b. For the devices annealed by ROA, only a weak dependence of base current and surface recombination velocity on annealing temperature is observed. In contrast, low temperature furnace annealing results in a strong reduction of J_{BO} and S with decreasing X_{jE}.

These results can be correlated with cross-sectional transmission electron microscopy (XTEM) of the poly/mono-Si interface. As an example, in the Figs.3.3a,b, XTEM cross sections of devices processed at 1100°C (ROA) and 900°C (furnace anneal) are compared. Though both anneals yield a similar emitter junction depth of about 70 nm, distinct differences are seen at the interface. Although for both samples the same HF-dip treatment was applied, a continuous oxide layer with $d_{ox} \sim 8$ Å is found in the case of the 900°C anneal. In contrast, the 1100°C anneal results in a breaking up of the "native" oxide layer into SiO_2 islands, leading to increased values of the surface recombination velocity as well as of the base saturation current.

3.1.3 Tunnelling Model

We explain the larger current gain that occurs in the case of a **continuous** oxide film in terms of tunnelling [3.6,7,9,10]. The larger bandgap of the insulating film forms a barrier to both electrons and holes. From the theoretical model of $\mathcal{H}u$ et al. [3.9], the effective surface recombination velocity representing both the interface, as well as the bulk properties of the polysilicon layer, is given by

$$S = S_i + \left[1/T_i + 1/(S_i + S_p)\right]^{-1} . \tag{3.2}$$

In (3.2) S_i represents recombination at surface states, T_i is the contribution due to the tunnelling of holes through the interfacial layer and S_p describes recombination of holes in the polysilicon layer. By varying the thickness of the polysilicon layer, recent experimental results have shown [3.16,7] that at least in the case of continuous interfacial oxide layers, S_p is not the crucial parameter for calculating S. In our calculations we assume $S_p = 1 \cdot 10^5$ cm/s.

Using the WKB approximation, we obtain for the tunnelling of holes through a rectangular barrier [3.17,9]

$$T_i = \left(\frac{kT}{2\pi m_h^*}\right)^{1/2} \frac{e^{-b_h}}{1 - C_h kT} \tag{3.3}$$

with

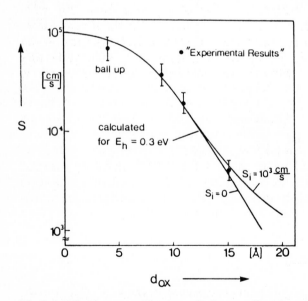

Fig.3.4. Effective recombination velocity as a function of oxide layer thickness

$$b_h = \frac{4\pi d_{ox}}{h} (2m_p^* E_h)^{1/2}$$

$$c_h = \frac{2\pi d_{ox}}{h} \left[\frac{2m_p^*}{E_h}\right]^{1/2}$$

where E_h is the height of the potential barrier and m_h^* is the effective mass for holes. In order to calculate S as a function of the oxide layer thickness, a value for E_h is required. As shown in Fig.3.4, a best fit to the "experimental" results for S - as obtained from measured base currents and XTEM studies - is obtained for $E_h \sim 0.3$ eV. We conclude that for the case of continuous interfacial oxide layers with $d_{ox} \gtrsim 10$ Å, the tunnelling model adequately predicts reduction in base current with increasing oxide-layer thickness.

3.1.4 Emitter Resistance

Interfacial oxide layers not only act as a barrier to hole transport, but also increase the resistance for electrons [3.18, 7]. Since the largest current density in the device passes through the emitter, specific emitter series resistance r_E must be kept as low as possible to prevent a degradation in transconductance [3.8]. We have used *Ning's* method [3.19] to determine the emitter resistance. Values of $r_E = 25\text{-}30$ $\Omega \cdot \mu m^2$ have been obtained for the devices annealed by ROA. No significant variations of r_E are observed for the different ROA anneals whereas, in the case of furnace annealing, a significant increase in r_E with decreasing X_{jE} is found (Fig.3.5). This increase in specific emitter resistance is

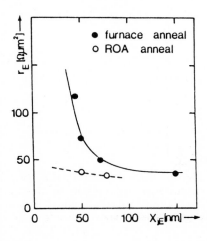

Fig.3.5. Specific emitter resistances r_E for a range of emitter drive in temperatures

caused by both an increase in the interface resistance and also, due to the lowered electrically active dopant concentration in the polysilicon layer, by an increase in the contact resistance [3.20].

As far as only current gain is concerned, furnace annealing would be of more practical use for the fabrication of polysilicon emitter transistors than ROA. However, because of the increase in contact, sheet and interface resistance at low-temperature furnace annealing, ROA is more promising for the fabrication of polysilicon emitter transistors with emitter junction depths below 50 nm.

3.2 Base Charge Control

To improve device speed, an important issue is the reduction of the base width W_B. Several problems are encountered with conventional boron ion implantation if W_B values below approximately 200 nm are desired. Due to the strong channelling effect for boron ions, either very low implantation energies (<10 keV) have to be chosen, or a thick screen oxide (>100 nm) has to be used for the active base implant. In either case control of base charge (and thus current gain) is extremely difficult, essentially because the peak of the boron distribution lies very close to, or even above, the silicon surface (Fig.3.6). Base charge, therefore, is highly sensitive to variations in emitter junction depth. A desirable situation for good process control is depicted in Fig.3.7a, where the emitter profile intersects the base implant profile at its peak.

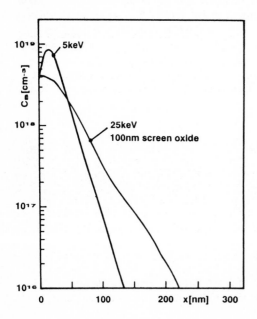

Fig.3.6. Boron profiles (SIMS) obtained after implantation at 5 keV (no screen oxide) and 25 keV (1009 nm screen oxide)

Figure 3.7b shows the base collector junction depths and minimum base widths W_B achievable under this condition. Since channelling is not properly described by current process simulation tools, an experimental study down to implantation energies of 2 keV has been done to obtain these data.

To avoid the detrimental boron channelling, preamorphization [3.21] of the single-crystal silicon with, e.g., a germanium implant prior to the active base implant, is a promising technique. We implanted n-type (100) Si wafers (resistivity 0.3Ω·cm) with $2 \cdot 10^{14}$ Ge/cm^2 at 40 keV. Into the now completely amorphous surface layer of 40 nm thickness, boron was implanted with a dose of $3 \cdot 10^{13}$ B/cm^2 at 10

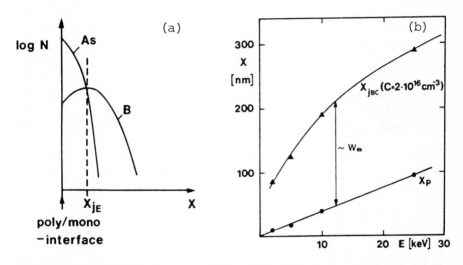

Fig.3.7a,b. (a) Schematic plot of As and B profiles in mono-Si using poly-Si as diffusion source for As. (b) peak boron penetration depth X_P and base-collector junction depth X_{jBC} at $2 \cdot 10^{16}$ cm^{-3}

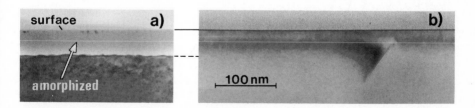

Fig.3.8a,b. XTEM micrograph showing amorphous zone obtained by implantation of Si with $2 \cdot 10^{14}$ Ge/cm^2 at 40 keV and annealed at 450°C (30min) (a). After an additional anneal at 1050°C (10s) dislocation loops are still observed (b). At 1100°C (10s) no defects are detectable any more (not shown)

51

keV. After depositing an oxide of 50 nm thickness at 400°C, the samples were annealed at 450°C for 30 minutes in Ar atmosphere. At this temperature a sharpening up of the amorphous/crystalline (α/X) interface occurs [3.22]. The XTEM micrograph of Fig.3.8a exhibits the amorphous zone. As the α/X interface is very smooth (peak to valley distance is less than 5nm) a necessary condition for perfect annihilation of defects is fulfilled. Figure 3.9 shows the thus obtained boron concentraion profile. The boron penetration depth X_{jBC} at 2.10^{16} B/cm^3 is only 113 nm, compared to X_{jBC} = 190 nm of an equivalent, but non-amorphized, sample. To recrystallize the amorphous zone, a high-temperature rapid optical anneal at 1050°C for 10 s in Ar atmosphere was performed. Although crystallization occurred, dislocation loops centered around the original α/X interface are still observed in XTEM micrographs (Fig.3.8b). Further increase of the temperature to 1100°C leads to the desired defect-free sample - as evidenced by a thorough XTEM investigation. This encouraging result is, however, accompanied by a broadening of the boron distribution. The evolution of the boron profile upon heat treatment is depicted in Fig.3.9.

These results indicate that further efforts are still necessary. To reduce the temperature budget, special emphasis will have to be paid to the optimization of the annealing cycles. Additionally, reducing the

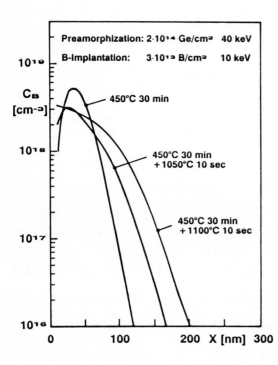

Fig.3.9. Boron profiles (SIMS) from implantations into pre-amorphized Si

energy for the boron implantation will place the boron peak well inside the amorphized layer and will contribute to a more complete elimination of chanelling. The final verdict concerning the usefulness of this technique will depend on the achievable device yield obtained in a full production run.

Another attractive method for achieving narrow base widths is double diffusion of boron and arsenic out of the emitter poly-Si layer, primarily because in this case the reference plane for both diffusion profiles is exactly the same. Unfortunately, it is not possible to perform both diffusion processes within a single heat cycle, because boron is nearly immobile in the presence of high arsenic concentrations. This holds even for a relatively long, high-temperature heat treatment at 950°C (30min). In this case the faster diffusing boron atoms should have the best chance to 'escape' the arsenic diffusion front (Fig.3.10a).

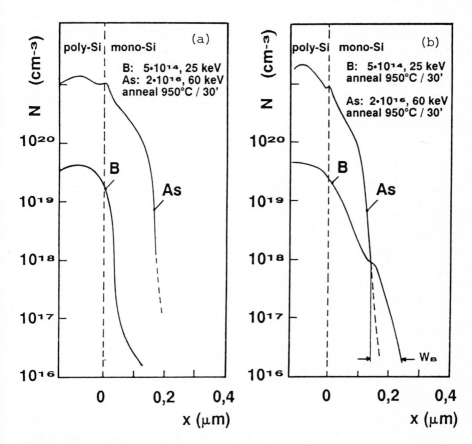

Fig.3.10a,b. Results of joint (a) and successive (b) diffusion of boron and arsenic from a single poly-Si layer

Successive diffusion of the two dopants, as shown in Fig.3.10b solves this problem and indeed yields very narrow base widths, especially at reduced heat cycles.

However, another problem is evident from Fig.3.10: Although the arsenic has 'seen' the same heat cycle in case a) and b), the As doping profiles obtained are quite different. This is because in case b) the poly-Si layer has already partially recrystallized during the first boron drive-in and thus As diffusion is slowed down [3.23] leading to the shallower profile with steep slope also in the poly-Si seen in Fig.3.10b. Thus, boron and arsenic profiles are not only coupled by the usual diffusion interaction [3.24], but also via changes in the poly-Si or the poly/mono interface morphology. A very tight control of the poly/mono-Si interface as well as of the heat cycles employed is therefore required in order to achieve reproducible, narrow base profiles by the double diffusion method.

Nevertheless, emitter junction depths of around 30 nm and base widths below 50 nm will certainly be controllable using some of the techniques discussed above or related methods (Chap.2). At this point we move on to consider limitations dictated by device physics.

3.3 Transit Time Considerations

One of the most important factors determining the speed of bipolar circuits is the transit time τ_F, which in first-order calculations is proportional to the square of the base width W_B. In reducing the base width in bipolar transistors there are some physical constraints to be considered. Since punchthrough between emitter and collector must be avoided at normal operating voltages (we assume $U_{EC} \leq 3V$), a reduction of base width W_B must be accompanied by a corresponding increase in maximum base doping concentration N_A. However, a limit is set by trap-assisted forward tunnelling in the E-B junction, which leads to strongly nonideal behaviour of the base current at large volues of N_A. Figure 3.11 shows base current characteristics for N_A ranging from $1 \cdot 10^{18} \text{cm}^{-3}$ to $3 \cdot 10^{19} \text{cm}^{-3}$. Although the trap density may depend somewhat on processing details, we assume a limit of $N_A = 1 \cdot 10^{19} \text{cm}^{-3}$ for proper transistor operation.

Under the restrictions of punchthrough and trap-assisted forward tunneling, we have determined the forward transit time τ_F by numerical simulation for various values of W_B and N_A. Since we are interested in the total transit time as well as in the separate contributions of the emitter and base charges to τ_F, we used quasistatic modelling. In this model cut-off frequency f_T is given by [3.25]

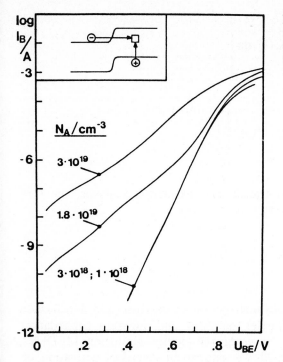

Fig.3.11. Base current characteristics for transistors with different base doping concentrations

$$\frac{1}{2\pi f_T} = \frac{dQ}{dI_c} \tag{3.4}$$

where I_C is the collector current and $Q = q\int p(x)dx$ is the hole charge integrated over the whole transistor. In (3.4) dQ/dI_C contains the contributions of the minority carrier diffusion charges Q_E, Q_B and Q_C stored in the emitter, base and collector as well as the junction capacitances C_{jEB} and C_{jBC}, resulting in

$$\frac{dQ}{dI_C} = \tau_E + \tau_B + \tau_C + \tau_{jEB} + \tau_{jBC} \tag{3.5}$$

with

$$\tau_E = \frac{dQ_E}{dI_C} \, , \, \tau_B = \frac{dQ_B}{dI_C} \, , \, \tau_C = \frac{dQ_C}{dI_C}$$

and

$$\tau_{jEB} = \frac{d}{dI_C} \left[\int_0^{U_{BE}} C_{jEB}(U) du \right] ,$$

$$\tau_{jBC} = \frac{d}{dI_C} \left[\int_0^{U_{BC}} C_{jBC}(U) dU \right] .$$

To be consistent with usual transit time extraction schemes, the forward transit time is defined by

$$\tau_F = \frac{1}{2\pi f_T} - \tau_{jEB} - \tau_{jBC} . \tag{3.6}$$

Using a Gaussian doping distribution in the base and a truncated Gaussian in the emitter, carrier densities and junction capacitances were calculated at $U_{BC} = 0$ using the device simulator MEDUSA [3.12]. Unless otherwise stated we assume an electrical activation of $N_D^+ = 7 \cdot 10^{19}$ cm^{-3} in the emitter and perform the simulations for an emitter junction depth of 40 nm and a surface recombination velocity of $7 \cdot 10^4$ cm/s. Emitter charge storage in the polysilicon layer is neglected in our calculations. This is justified since recent experiments

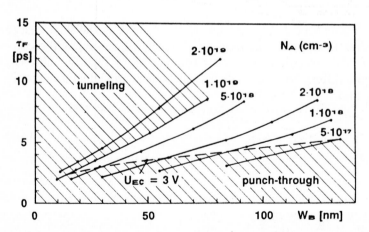

Fig.3.12. Transit time τ_F vs base width W_B for different values of the maximum base doping concentration N_A. The shaded areas indicate E-B forward tunnelling and E-C punchtrough

56

[3.26] have shown that in devices with small X_{jE} base current is independent of the polysilicon layer thickness for $d_{poly} \geq 10$ nm. This even
holds if the devices are annealed by ROA below 1100°C.

Figure 3.12 shows τ_F vs W_B for different N_A values. The shaded
areas indicate E-B forward tunnelling and E-C punchthrough, as discussed above, allowing a transit time τ_F of approximately 3 ps at W_B =
20 nm. As expected, τ_F decreases more than linearly with decreasing
W_A. On the other hand, if one increases the doping concentration N_A
in order to achieve a low base sheet resistance and to suppress punch
through, τ_F increases. To investigate the origin of this behaviour in
more detail, we have extracted τ_E and τ_B separately. Figure 3.13 shows
τ_F, τ_E and τ_B vs N_A for W_B = 50 nm. Due to the reduction in electron mobility τ_B first increases with N_A. At higher base doping
concentrations the electron mobility begins to saturate and the internal
drift-field increases, resulting in nearly constant values for τ_B. In contrast, τ_E increases with $1/I_C$, and at doping levels $N_A > 5 \cdot 10^{18}$ cm^{-3}
the change in total transit time is mainly due to the increase in τ_E, indicating the importance of emitter charge storage for shallow devices.

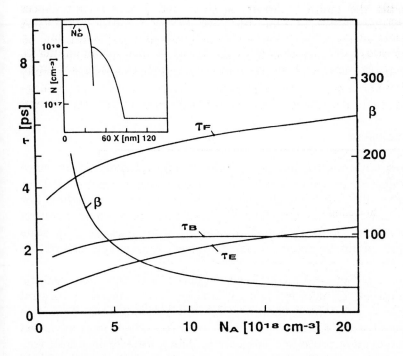

Fig.3.13. Forward transit times τ_E, τ_B, τ_F and current gain β as a function of base
doping level N_A. The base width (W_B = 50nm) and emitter profile (N_D^+ = $7\cdot10^{19}$
cm^{-3}, X_{jE} = 40nm, S = $7\cdot10^4$cm/s) were kept fixed during the calculations

57

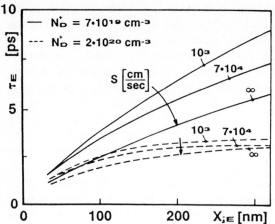

Fig.3.14. Emitter transit time τ_E as a function of emitter junction depth for various values of electrical activation and surface recombination velocity

In Fig.3.14 τ_E is plotted as a function of emitter depth for various values of S and electrical activation $N_D{}^+$. The base profile (W_B = 50 nm, N_A = $5 \cdot 10^{18}$ cm^{-3}) and impurity gradient of the emitter was kept fixed during these calculations.

From our results in Fig.3.13,14 we conclude that for optimum performance of downscaled transistors, minimization of both τ_E and τ_B is of importance. To decrease minority carrier storage in the emitter as far as possible very shallow junctions and a high electrical activation are needed.

3.4 Conclusion

Transistors with emitter junction depths of approximately 50 nm have been fabricated. Particularly if rapid annealing techniques are used for the emitter drive-in, excellent device characteristics - including low emitter resistance - are achieved. We have further shown that several methods relying on conventional processing techniques exist to scale down base widths to about 50 nm. However, considerable efforts are still necessary to improve process control and reliability.

Based on these technological parameters and assuming realistic doping profiles, an improvement of intrinsic device speed to forward transit times of 3-4 ps is deduced from device simulations. This is to be compared with figures of about 20-30 ps for today's most advanced bipolar production processes. The corresponding increase in transit frequency up to about 50 GHz gives an indication of the speed achievable for small scale silicon bipolar circuits on the basis of well-proven processing techniques.

Since the emitter charge storage contribution to τ_F amounts to about 30% for these extremely shallow devices, further performance improvement may be expected by employing true heterojunction devices such as SiC-emitter transistors.

Acknowledgement. We would like to thank Dr. H. Cerva and Miss C. Fruth for the TEM work.

References

3.1   A.W. Wieder: IEDM Techn. Dig. 8 (1986)
3.2   T.H. Ning, R.D. Isaac: IEEE Trans. ED-27, 2051 (1980)
3.3   H.C. de Graaf, J.G. de Groot: IEEE Trans. ED-26, 1771 (1979)
3.4   J. Graul, A. Glasl, H. Murrmann: IEEE J. SC11, 491 (1976)
3.5   Chung C. Ng, Edward S. Yang: IEDM Techn. Dig. 32 (1986)
3.6   G.R. Wolstenholme, N. Jorgensen, P. Ashburn, G.R. Booker: J. Appl. Phys. 61, 225 (1986)
3.7   B. Benna, T.F. Meister, H. Schaber: Solid-State Electron. 30, 1153 (1987)
3.8   J.M.C. Stork, E. Ganin, J.D. Cressler, G.L. Patton, S.A. Sai-Halasz: IBM J. Res. Develop. 31, 617 (1987)
3.9   Z. Yu, B. Ricco, R.W. Dutton: IEEE Trans. ED-31, 773 (1984)
3.10 A.A. Eltoukly, D.J. Roulston: IEEE Trans. ED-29, 1862 (1982)
3.11 B. Benna, T.F. Meister, H. Schaber, A. Wieder: IEDM Techn. Dig. 302 (1985)
3.12 W.L. Engl: "MEDUSA User Manual", ITHE Aachen, Techn. Hochschule Aachen (1985)
3.13 J.W. Slotboom, H.C. De Graaf: Solid-State Electron. 19, 857 (1976)
3.14 J. Dziewior, W. Schmid: Appl. Phys. Lett. 31, 346 (1977)
3.15 N.D. Arora, J.R. Hauser, D.J. Roulston: IEEE Trans. ED-29, 292 (1982)
3.16 G.L. Patton, J.C. Bravmann, J.D. Plummer: IEEE Trans. ED-33, 1754 (1986)
3.17 R. Stratton: J. Phys. Chem. Solids 23, 1177 (1962)
3.18 J.M.C. Stork, C.Y. Wong, M. Arienzo: Symp. on VLSI Technology (May 1985), Dig. Techn. Papers, p.54
3.19 T.H. Ning, D.D. Tang: IEEE Trans. ED-31, 409 (1984)
3.20 H.J. Böhm, H. Kabza, T.F. Meister, H.J. Wendt: Spring Meeting Electrochem. Soc., Philadelphia (May 1987) p.347
3.21 D.K. Sadana, et al.: Electrochem. Soc. 131, 943 (1984)
3.22 G.A. Rozgonyi, et al.: In *Semiconductor Silicon 1986*, ed. by H. Huff (Electrochemical Society, Pennington, N.J. 1986) p.696
3.23 V. Probst, et al.: In [Ref.3.22, p.594]
3.24 A.F.W. Willoughby: In *Impurity Doping Processes in Silicon*, ed. by F.F.Y. Wang (North-Holland, Amsterdam 1981) p.1
3.25 L.J. Varnerin: Proc. IRE. 47, 523 (1959)
3.26 T.F. Meister et al., to be published

4. Trench Isolation Schemes for Bipolar Devices: Benefits and Limiting Aspects

Hiroshi Goto and Katsuyuki Inayoshi

Bipolar Process Division, Fujitsu Limited
1015, Kamikodanaka, Nakahara, Kawasaki, 211, Japan

This chapter gives a review of benefits and limiting aspects of the trench isolation techniques for bipolar devices. The most sophisticated trench isolation techniques have realized not only a high packing density but also reduced collector-substrate, wiring-substrate and base-collector parasitic capacitances. By using these techniques, high performance bipolar devices such as ultra-high-speed ECL RAMs, gate arrays and microprocessors have been fabricated. However, crystalline defects caused by trench structures continue to pose serious problems. Trench isolation techniques are still in the process of development, and it seems that there is no apparent limiting aspect unless the trench width exceeds the filler material width necessary to sustain enough breakdown voltage.

4.1 Background

Isolation is one of the key technologies in integrated circuit processes. When the Integrated Circuit (IC) was invented by J.S. Kilby, each electrical element of the device was isolated by slots and air. However, with the advent of the planar technique, pn-junction isolation became the major technique because Kilby's nonplanar structure was unfavourable for wiring processes. Another approach was also tried to improve the surface topography of air-isolated devices. This is the so-called dielectric isolation utilizing groove etching. Air-isolation slots were replaced by dielectric substrates in which active semiconductor islands were sustained. Dielectric isolation achieved rather planar surfaces, but it was too complicated for conventional devices. Only for very high voltage devices has this technique been used.

After the LOCOS technique was developed, pn-junction isolation was gradually replaced by oxide isolation. At the same time, there was a revival of groove isolation. In this case, only laterally isolated regions were formed by grooves filled with dielectric materials and the substrate was isolated by a pn-junction. Grooves were formed by aniso-tropic chemical etching. Both of these isolation techniques achieved higher packing density than pn-junction-isolated devices. The scaling concept was also employed to achieve large-scale integration and high-performance devices. When the devices are scaled, not only active regions but also inactive regions such as the isolation should be reduced. In the case of LOCOS, the so-called bird's beak enlarges the inactive regions. The length of bird's beak is dependent on the thick-ness of oxide and cannot be reduced proportionally to the scaling ratio. This was the ultimate limitation of oxide isolation. On the other hand, groove isolation utilized anisotropically etched V-shaped grooves for isolation using alkaline etchant. The V-groove's depth is 70% of its width, so the reduction of isolation regions is limited by the groove depth necessary to keep enough breakdown voltage.

In order to overcome these problems of conventional isolation techniques, the bird's beak free and width-independent isolation scheme was most important, the result was the advent of trench isola-tion. This is a new technology employing novel etching techniques such as RIE to form narrow and steep U-shaped grooves, or trenches for isolation, but its root is based on the classical idea of Kilby's in-vention or dielectric isolation.

A practical trench isolation technique was first reported in 1978 [4.1] and its application to devices appeared early in the 1980s [4.2,3]. Since the first application of trench isolation to 1kb ECL RAMs in 1982 [4.2], many bipolar devices have been fabricated by various kinds of trench isolation techniques. In the case of bipolar devices, some trench isolation techniques have been used in mass production since the early stage of the development, though trench isolation in MOS devices is only in experimental trial stages. This is partly because isola-tion in bipolar devices plays a more important role in their perfor-mance than in MOS devices.

In the following sections, process technologies, device structures, applications to devices and benefits and limiting aspects of trench iso-lation for bipolar devices will be discussed.

4.2 Process Technologies and Device Structures (First Generation)

Typical trench isolation techniques [4.1-3] consist of three major process steps. These are trench etching, trench filling and planarization. Figure 4.1 shows an example of process steps. As usual, an n^+ buried

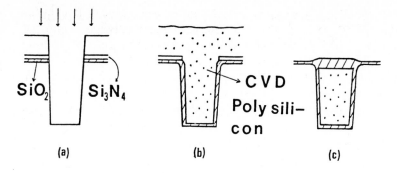

Fig.4.1. Fabrication process steps of a typical trench isolation technique

layer and an n^- epitaxial layer are formed on a p-type substrate. On the n^- epitaxial layer, a nitride layer with thin pad oxide is deposited and a masking layer for trench etching is also deposited. After the isolation pattern is formed, trench etching follows using anisotropic dry etching such as RIE. The trenches penetrate both the epitaxial and buried layers to reach the p-type substrate. The inside of the trenches is then oxidized to accomplish electrical isolation. After that, trenches are buried with undoped polysilicon. Excess polysilicon is polished off until the nitride layer appears. Finally the surface of polysilicon is oxidized for capping. The trenches are filled with oxide and polysilicon.

Trench etching is one of the most important processes in trench isolation techniques. The typical structure of narrow and deep trenches for isolation is achieved by anisotropic plasma etching. RIE is widely used for this purpose. The etching conditions should be carefully optimized to avoid unfavourable trench shapes. Figure 4.2 shows examples of maladapted trench structures. Figure 4.2a shows so-called "black silicon" or needle-like silicon residues; and "trenching" at the corners of bottom edges of trenches is shown in Fig.4.2b. Figure 4.2c displays the case of lateral etching that occurs at the highly doped n^+ buried layer. All of these unfavourable shapes cause large leakage current between isolated elements. Figure 4.2d shows the "barrel-like shape" or under-cutting structure. In this case the trenches are not completely buried by filler material, and voids occur in isolation regions. These voids may become sharp crevices after the planarization step has been completed. They cause bad metallization step coverage.

These trench shapes are strongly dependent on gas species, pressure, flow rates and power in the reactive-ion-etching processes [4.4,5]. These etching conditions are not universal, so they should be optimized according to etching systems. Figure 4.3 shows a cross-sec-

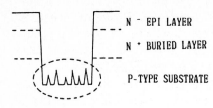

N⁻ EPI LAYER

N⁺ BURIED LAYER

P-TYPE SUBSTRATE

(a) Black silicon

Fig.4.2. Unfavourable trench shapes caused by trench etching

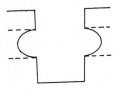

(b) Trenching

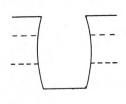

(c) Lateral etching

(d) Barrel-like shape

Fig.4.3. Cross-sectional SEM micrograph of an ideal trench shape

tional SEM micrograph of an ideal trench shape. There is no black silicon, no trenching, no lateral etching and no under-cutting.

Trench filling and planarization processes are important; the trenches should be filled completely with filler material. If not, non-planar surfaces occur after excess filler material is removed. The filler material should also be deposited as uniformly as possible. There are a variety of possible filler materials such as CVD oxide, CVD nitride and undoped polysilicon. When CVD oxide is used, trench shapes, especially the sidewall taper should be optimized because CVD oxide does not show good step coverage inside trenches [4.1]. Recently, instead of CVD oxide, bias sputtered oxide, ECR plasma enhanced CVD oxide have been tried. Nevertheless, low pressure CVD polysilicon shows excellent step coverage although it is not a complete insulator. Undoped polysilicon is widely used as the filler material. CVD nitride is a good insulator, but because of its high tension it is impossible to fill trenches completely. Nitride is often used as a spacer between thermal oxide of trench surfaces and undoped polysilicon. Excess filler material is removed to attain a planar surface by controlled etch-back techniques or polishing techniques. This planarization process requires high uniformity. In our experience, polishing techniques are superior to controlled etch-back techniques.

Figure 4.4 shows the typical trench-isolated structures. Figure 4.4a illustrates Deep Groove Isolation (DGI) [4.1]. Trenches are etched by RIE using Cl_2/Ar gas mixtures. Trench surfaces are oxidized and filled with CVD oxide. Excess oxide is removed by a controlled etch-back technique. Figure 4.4b displays Isolation by Oxide and Polysilicon; second version (IOP-II) [4.2]. Trenches are etched by RIE and filled with oxide and polysilicon. Excess polysilicon is removed by a polishing technique. IOP-II is an advanced version of conventional V-groove IOP, and only V-groove etching was replaced by trench etching. Figure 4.4c exhibits U-groove Isolation (U-Iso) [4.3]. In this case, the trench is not U-shaped but Y-shaped. The trench etching process consists of two steps. First, anisotropic chemical etching of (100) surface is employed to form an overhung structure using a nitride layer with side-etched pad oxide as a mask. Then anisotropic dry etching such as reactive sputter etching using a CCl_4/O_2 gas mixture follows for trenches to penetrate an n^+ buried layer. The inside of the trenches is oxidized, and the second nitride layer is deposited to cover the trench surface. Trenches are filled with undoped

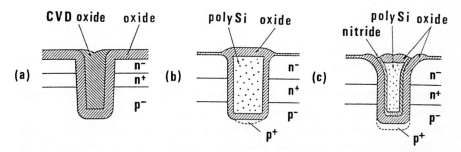

Fig.4.4a-c. Typical trench-isolated structures. (a) DGI [4.1], (b) IOP-II [4.2], (c) U-Iso [4.3]

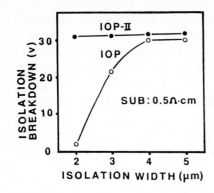

Fig.4.5. Isolation breakdown voltage vs. groove width. The epi-thickness is $1.5\mu m$. (comparison of V-groove IOP and U-groove IOP-II)

polysilicon. Excess polysilicon is removed by a controlled etch-back technique.

The electrical characteristics of the isolation breakdown voltages are illustrated with Fig.4.5, comparing IOP-II with conventional V-groove IOP. In the case of V-groove IOP, 2 μm deep grooves cannot isolate 1.5 μm thick epitaxial layers, while U-grooves or trenches can isolate electrical elements independently of their widths [4.2]. In recent experiments it has been found that even if the trench width is in the submicron range, sufficient breakdown voltages are obtained.

These trench isolated structures may be called the first generation. In these trench isolation techniques, one attempts to reduce only the isolation regions; other inactive regions such as field regions for wiring or shallow isolation for base and collector-reach-through regions are not taken into account. As far as memories are concerned, there is no serious problem. However, for high-performance logic devices, those parasitic inactive regions are fatal disadvantages for switching speeds. In order to overcome these drawbacks of the first generation, some process modifications followed. In the next section, process technologies and device structures of advanced trench isolation techniques, which may be called the second generation, will be described.

4.3 Process Technologies and Device Structures (Second Generation)

Process modifications for the second generation may be roughly divided into two types. One is the combination of LOCOS or ROX (Recessed OXide) with deep trenches, and the other is the combination of shallow grooves or moats with deep trenches. For instance, Fig.4.6 shows the process steps of U-FOX (U-groove isolation with thick Field OXide, renamed from IOP-L) or an advanced version of IOP-II,

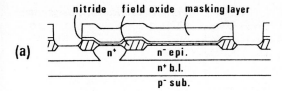

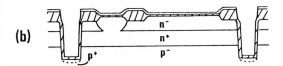

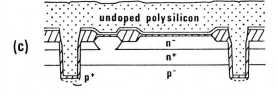

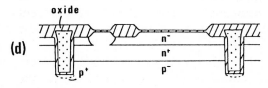

nitride   field oxide   masking layer

(a) n^+ n^- epi.
       n^+ b.l.
       p^- sub.

(b) p^+ n^-
                 n^+
                 p^-

undoped polysilicon

(c) p^+ n^-
                 n^+
                 p^-

oxide

(d) p^+ n^-
                 n^+
                 p^-

Fig.4.6. Fabrication process steps of U-FOX [4.6]

combining LOCOS with trench isolation [4.6]. After the formation of an n^- epitaxial layer by conventional means, its surface is selectively oxidized except for base, collector-reach-through regions and isolation regions, using the LOCOS technique. Then a new nitride layer and a masking layer for trench etching are deposited. They are removed at the isolation regions and trench etching follows. After the trenches have been etched, their surfaces are oxidized and filled with undoped polysilicon. Excess polysilicon is polished off until the polysilicon not on the isolation regions is completely removed. Then capping oxidation is performed.

Another approach is shown in Fig.4.7. In this technique [4.7], deep trenches for device isolation are first etched, using a thick oxide layer as a mask. Then the masking layer is selectively removed by the second isolation patterns, which contain base- and collector-reach-through isolation and field regions. Again trench etching is performed to give no penetrating shallow grooves. Deep trenches penetrate the n^+ buried layer to reach also the p-type substrate. Both deep trenches and shallow grooves are buried with CVD oxide. The CVD oxide itself

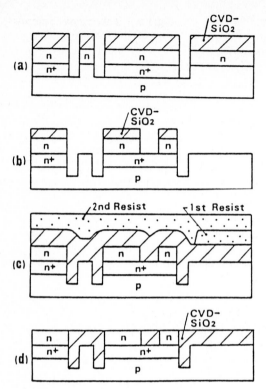

Fig.4.7. Fabrication process steps [4.7]

does not achieve good step coverage in the trenches, but by optimizing the distances between deep trenches in shallow grooves, voids in the deep trenches are elaborately covered by over-grown oxide. Excess oxide is removed by a controlled etch-back technique in just the same way as reported for simplified BOX isolation for MOS devices [4.8].

There are some structural modifications of these techniques. Figure 4.8 shows structural variations of modified trench-isolated devices in the second generation. Figure 4.8a is an example of deep trench isolation combined with ROX. The recessed oxide (ROX) is formed by recessing silicon using RIE followed by oxidation. Deep trenches are then etched, covered with a thin thermal oxide layer and filled with CVD oxide. In some cases there are voids in the middle of isolation regions [4.9,10]. Figure 4.8b is an example of a dual-depth trench-isolated structure [4.11]. Shallower trenches, filled with oxide, nitride, and polysilicon isolate transistors and Schottky barrier diodes and they may be able to isolate base and collector-reach-through regions. Figure 4.8c shows the U-FOX structure. LOCOS is used for base and collector-reach-through isolation and field oxide for wiring regions. Figure 4.8d is the final structure of the technique mentioned

68

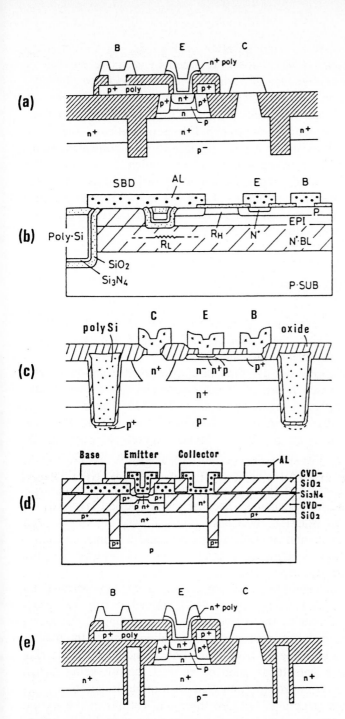

Fig.4.8a-e. Modified trench-isolated structures. (a) [4.9], (b) [4.11], (c) [4.6], (d) [4.7], (e) [4.12]

above [4.7]. Figure 4.8e is a modification of Fig.4.8a. As a filler material for deep trenches, CVD oxide is replaced by selective epi-silicon [4.12,13]. In this case, trenches of variable width can be filled and planarized at the same process step. There are other variations on advanced trench-isolated structures [4.14,15]. In [4.14], field oxidation, as well as capping oxidation of trench-filling polysilicon, was performed after excess polysilicon has been removed by a controlled etchback technique. In order to bury the crevices which occur along the trench-edge surface, an etch-back technique of reflow glass was used. The final device structure resembles Fig.4.8d. In [4.15] vertical pnp transistors were formed as well as conventional npn transistors; a new advantageous device structure was achieved.

4.4 Benefits and Device Performance

The trench isolation techniques mentioned above have a number of advantages and by using them bipolar devices have achieved higher performance than conventionally isolated devices.

First, higher packing density than in conventional isolation is achieved because the trench depth is almost independent of its width and the bird's beak is reduced much less than in LOCOS. The p^+ channel stop layers are usually formed at the bottom of trenches, so they are completely separated from n^+ buried layers. Therefore, even if the trench width is reduced, there are enough spaces that a depletion layer can sustain high breakdown voltage. Also, the complete separation of the channel stop layers and the n^+ buried layers makes collector-substrate parasitic capacitances smaller. Because the n^+ buried layer islands are self-aligned to isolation trenches, there is no alignment step between a buried layer and an isolation pattern and a planar collector-substrate pn-junction is achieved, which is also favourable for the reduction of its parasitic capacitance.

Figure 4.9 is an example of device performance demonstrated by 1kb ECL RAMs scaled using IOP-II [4.2]. Trenches can be scaled down by the same factor as the active regions. Accordingly, the access time can be reduced in proportion to the scaling factor. In another example [4.3], the advantage of trench isolation is shown by using ring oscillators. The basic gate delay time of an ECL circuit using trench isolation is nearly 30% improved compared with oxide isolation.

Not only trench structures but also the modifications in the second generation have much improved the device performance. By introducing thick LOCOS oxide or oxide-filled shallow grooves for wiring

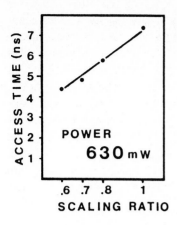

Fig.4.9. Address access time of 1kb ECL RAMs versus scaling ratio

regions, parasitic capacitances between wiring layers and substrate have been reduced. This approach is suitable for logic devices such as gate arrays. As well as the reduction of wiring-substrate capacitances, base-collector parasitic capacitances are also reduced because LOCOS or shallow grooves surround the base regions to achieve a so-called walled-base structure. In the case of the U-FOX structure, the wiring delay time of 50 ps/mm is achieved for the first and second metallization layers using a 1.6 μm wide wiring pattern. A loaded gate delay time (FI = FO = 3, wiring length = 2 mm, 0.2 mA/gate) of 325 ps/gate using an ECL circuit is obtained [4.6]

Recently, much improved results have been reported using these trench-isolated structures and sophisticated base-emitter self-aligned processes. Table 4.1 shows the results of basic gate delays using ring oscillators of ECL or CML circuits. Sub-100 ps delays are achieved, and even a sub-40 ps delay is obtained using the U-FOX structure combined with a base-emitter self-aligned process. The advantage of trench isolation compared with oxide isolation is also demonstrated in a self-aligned process named SDD [4.7]. By using trench isolation, a nearly 20% improvement of basic delay is achieved even if self-aligned transistors are employed.

4.5 Application to Practical Devices

Ring oscillators are convenient to compare device performance, but they are insufficient for investigating total performance of, e.g., access time, toggle frequency, cycle time and so on. Because of the many advantages and good results from ring oscillators, a lot of practical devices have been fabricated by trench isolation techniques. Table 4.2

71

Table 4.1. Ring oscillator results demonstrated by trench isolation techniques combined with sophisticated self-aligned transistors

Laboratory	Circuit	Performance	Process	Ref.
Fairchild	ECL	80ps/gate at 0.9mA/gate	$1.5 \times 2.5 \mu m^2$ emitter $f_\tau = 10 GHz$	4.14
IBM	ECL	73ps/gate at 12mW/gate	$0.8 \times 2.0 \mu m^2$ emitter	4.12
Matsushita	CML	53ps/gate at 3.2mW/gate	$0.5 \times 12 \mu m^2$ emitter $f_\tau = 16 GHz$	4.87
Fujitsu	ECL	39ps/gate at 1.28mA/gate	$0.35 \times 10 \mu m^2$ emitter $f_\tau = 17 GHz$	4.16

shows the list of bipolar devices excluding ring oscillators fabricated by trench isolation. In the early stages, memory devices received most attention but later logic devices, not only gate arrays, but prescaler and microprocessor were demonstrated. In 1983 only two devices were reported at ISSCC, while in 1986 the number had risen to seven.

There are many types of trench isolation techniques that are applied to practical devices. In addition to the previously mentioned techniques, other trench isolation techniques are included in Table 4.2. IMOX-III-Slot is the trench isolation filled with oxide and polysilicon. ADB-III uses trenches filled with CVD oxide. Other trench isolation techniques named Impact-X [4.34] or ExCL [4.35] are also reported. They are, or will be, used in mass production although no actual device has been presented at a conference.

4.6 Problems and Limiting Aspects

Trench isolation is now becoming essential for bipolar devices and several companies are manufacturing devices in volume. However, we cannot say that trench isolation has yet progressed to a mature development stage. Some problems are still left unsolved. These may be broadly divided into two categories; the structural problems and the electrical problems. They are often strongly related and the former, in particular, is apt to influence the latter. As previously described in Sect.4.2, if the trench etching process is not optimized, trench shapes

Table 4.2. Bipolar devices using trench isolation techniques (excluding ring oscillators) (T_{aa}: address access time, T_{pd}: basic gate delay time, F_t: toggle frequency, T_c: cycle time, [a]: Trench filled with oxide and polysilicon, [b]: Dual depth trenches filled with oxide, nitride and polysilicon)

Laboratory	Device	Performance	Process	Ref.
Fujitsu	1Kb ECL RAM	T_{aa}=4.4ns		4.2
	4Kb ECL RAM	T_{aa}=3.5ns		4.17
	16Kb ECL RAM	T_{aa}=15ns		4.18
	64Kb ECL RAM	T_{aa}=10ns	IOP-II[a]	4.19
	16Kb ECL RAM &	T_{aa}=2.8ns &		4.20
	1.2Kb gate array	T_{pd}=280ps		
	64Kb ECL RAM	T_{aa}=5ns		4.21
	Prescaler	F_t=1.6GHz	U-FOX; LOCOS & IOP-II	4.22
Hitachi	4Kb ECL RAM	T_{aa}=2.5ns	U-groove[b]	4.23
	16Kb ECL EAM	T_{aa}=3.5ns		4.24
NEC	16Kb ECL RAM	T_{aa}=4ns	Trench (no details)	4.25
NTT	ECL 5K G.A.	T_{pd}=165ps	SST & Trench[a]	4.26
IBM	32Kb RAM	T_{aa}=3ns	Poly-Si filled Trench	4.27
	5Kb ECL RAM	T_{aa}=0.85ns	ROX & Selective epi-Si filled Trench	4.28
	32b processor	T_c=60ns	Deep Trench (no details)	4.29
AMD	128Kb PROM	T_{aa}=35ns	IMOX-III-Slot[a]	4.30
Fairchild	16x4 Reg.File	T_{aa}=0.9ns	SAPT[a]	4.31
Honeywell	PLA	T_{pd} x power =26fJ	ADB-III(trench filled with oxide)	4.32
Matsushita	8-bit A/D converter	200 Msps	A-SMASH[a]	4.33

may change variously. The structural problems without electrical problems are usually due to the bad step coverage of metallization layers caused by voids or sharp crevices on isolation regions. In order to avoid these problems trench shapes should not be barrel-like structures or overhung structures. Over-etching in etch-back processes to remove excess filler material should be carefully controlled. Other maladapted trench structures such as black silicon, trenching, lateral etching of highly doped buried layers often accompany crystalline defects. These crystalline defects cause large leakage current between isolated elements. Thus, trench etching processes should be optimized thoroughly to achieve the best conditions.

The most serious problem is how to get rid of isolation leakage caused by crystalline defects. Crystalline defects often occur at stress-concentrated places such as sharp-edged corners caused by unfavourable trench structures. But even if the trench shapes show the ideal structure, crystalline defects may sometimes appear. They seem to be influenced by heat cycles and/or ion implant dosage for device processes. The detailed mechanism of crystalline defects are still unknown although some process modifications have been tried. The use of Y-shaped grooves instead of U-shaped grooves is one way to avoid right-angled corners of top edges of trenches at the expense of packing density [4.3, 36, 37]. In relation to trench isolation combined with LOCOS, some process modifications have been also proposed. One is the nitride spacer between the oxide of trench surfaces and filled polysilicon [4.38], and another is the combination of SEPOX with trenches which is applied to a CMOS static RAM [4.39]. In both of these techniques, selective oxidation of field regions is performed at the same time as capping oxidation. Without these modifications large stresses occur at the top corners of trenches when single crystalline silicon and filled polysilicon are oxidized simultaneously, and they cause crystalline defects.

Even though some problems remain, as mentioned above, trench isolation techniques are still progressing, and it seems there is no limitation in sight. At present, trench widths are limited by lithographic systems, not by other factors such as etching or deposition. However, the width cannot be reduced to less than the coating or filler material width. Accordingly, the thickness of insulator needed for satisfactory electrical isolation limits the trench width. We may claim that air isolation is the ultimate goal of trench isolation [4.40]. But before reaching this point, other process obstacles such as metallization reliability must be removed, and these are the present limiting factors for scaling down.

4.7 Conclusion

Trench isolation is one of the key technologies in high-performance bipolar devices. It realizes higher packing density and smaller collector-substrate parasitic capacitances than conventional isolation techniques. It also reduces wiring-substrate and base-collector parasitic capacitances with process modifications.

Trench isolation, as well as the so-called sophisticated self-aligned-transistor structures, is leading bipolar technology towards ultra-large-scale and high-performance integrated circuits.

References

4.1   J.A. Bondur, H.B. Pogge: Method for forming isolated regions of silicon utilizing reactive ion etching, U.S. Patent 4104086, Aug. 1978

4.2   H. Goto, T. Takada, R. Abe, Y. Kawabe, K. Oami, M. Tanaka: An isolation technology for high performance bipolar memories-IOP-II. IEDM Techn. Dig. (1982) pp.58-61

4.3   A. Hayasaka, Y. Tamaki, M. Kawamura, K. Ogiue, S. Ohwaki: U-groove isolation technique for high speed bipolar VLSIs. IEDM Techn. Dig. (1982) pp.62-65

4.4   G.C. Schwartz, P.M. Schaible: Reactive ion etching of silicon. J. Vac. Sci. Technol. 16, 410-413 (1979)

4.5   H.B. Pogge, J.A. Bondur, P.J. Burkhardt: Reactive ion etching of silicon with $Cl_2/Ar(l)$. J. Electrochem. Soc. 130, 1592-1597 (1983)

4.6   H. Goto, T. Takada, K. Nawata, Y. Kanai: A new isolation technology for bipolar VLSI logic (IOP-L). Symp. on VLSI Technology Techn. Dig. (1985) pp.42-43

4.7   H. Sakai, K. Kikuchi, S. Kameyama, M. Kajiyama, T.Komeda: A trench isolation technology for high-speed and low-power dissipation bipolar LSIs. Symp. on VLSI Technology, Techn. Dig. (1987) pp.17-18

4.8   T. Shibata, R. Nakayama, K. Kurosawa, S. Onga, M. Konaka, H. Iizuka: A simplified BOX (Buried OXide) isolation technology for megabit dynamic memories. IEDM Techn. Dig. (1983) pp.27-30

4.9   D.D. Tang, P.M. Solomon, T.H. Ning, R.D. Isaac, R.E. Burger: 1.25μm deepgroove-isolated self-aligned ECL circuits. ISSCC Techn. Dig. (1982) pp.242-243

4.10   S.F. Chu, G.R. Srinivasan, H. Bhatia, B.M. Kemlage, F. Barson, J. Mauer, J. Riseman: A self-aligned bipolar transistor. VLSI Science and Technology, Techn. Dig. (1982) pp.306-314

4.11   Y. Tamaki, T. Shiba, N. Honma, S. Mizuo, A. Hayasaka: New U-groove isolation technology for high-speed bipolar memory. Symp. on VLSI Technology, Techn. Dig. (1983) pp.24-25

4.12   D.D. Tang, G.P. Li, C.T. Chuang, D. Danner, M.B. Ketchen, J. Mauer, M. Smyth, M. Manny, J.D. Cressler, B. Ginsberg, E. Petrillo, T.H. Ning, C.C. Hu, H.S. Park: 73ps Si bipolar ECL circuits. ISSCC Techn. Dig. (1986) pp.104-105

4.13   G.P. Li, T.H. Ning, C.T. Chuang, M.B. Ketchen, D.D. Tang, J. Mauer: An advanced high-performance trench-isolated self-aligned bipolar technology. IEEE Trans. ED-34, 2246-2253 (1987)

4.14   M. Vora, Y.L. Ho, S. Bhamre, F. Chien, G. Bakker, H. Hingarh, C. Schmitz: A sub-100 picosecond bipolar ECL technology. IEDM Techn. Dig. (1985) pp.34-37

4.15   H. Sadamatsu, M. Inoue, A. Matsuzawa, A. Kanda, H. Shimoda: New self-aligned complementary bipolar transistors using selective-oxidation mask. IEDM Techn. Dig. (1984) pp.753-756

4.16 K. Ueno, H. Goto, E. Sugiyama, H. Tsunoi: A sub-40ps ECL circuit at a switching current of 1.28mA. IEDM Techn. Dig. (1987) pp.371-374

4.17 K. Ooami, M. Tanaka, Y. Sugo, R. Abe, T. Takada: A 3.5ns 4kb ECL RAM, ISSCC Techn. Dig. (1983) pp.114-115

4.18 K. Toyoda, M. Tanaka, H. Isogai, C. Ono, Y. Kawabe, H. Goto: A 15ns 16kb ECL RAM with a PNP load cell. ISSCC Techn. Dig. (1983) pp.108-109

4.19 Y. Okajima, K. Tokuda, K. Awaya, K. Tanaka, Y. Nakamura: 64kb ECL RAM with redundancy, ISSCC Techn. Dig. (1985) pp.48-49

4.20 Y. Sugo, M. Tanaka, Y. Mafune, T. Takeshima, S. Aihara, K. Tanaka: An ECL 2.8ns 16k RAM with 1.2k logic gate array. ISSCC Techn. Dig. (1986) pp.256-257

4.21 T. Awaya, K. Toyoda, O. Nomura, Y. Nakaya, K. Tanaka, H. Sugawara: A 5ns access time 64kb ECL RAM. ISSCC Techn. Dig. (1987) pp.130-131

4.22 H. Suzuki, T. Akiyama, K. Ueno: A 1.6GHz low power silicon dual modulus prescaler IC. IEDM Techn. Dig. (1984) pp.682-685

4.23 K. Yamaguchi, K. Kanetani, H. Todokoro, T. Nakano, K. Akimoto, K. Ogiue: An ECL 4k-bit bipolar RAM with an effective access time of 2.5ns and on-chip address latches. Symp. on VLSI Technology Techn. Dig. (1984) pp.52-53

4.24 K. Yamguchi, H. Nambu, K. Kanetani, N. Homma, Y. Nishioka, A. Uchida, K. Ogiue: A 3.5ns, 2W, 20mm^2 16kb ECL bipolar RAM. ISSCC Techn. Dig. (1986) pp.214-215

4.25 M. Arimura, M. Nakamae, T. Tashiro, S. Ohi, T. Kamiya, S. Kishi, Y. Minato, J. Nokubo, T. Tamura: A 4ns access time 4kx4 ECL RAM. ISSCC Techn. Dig. (1986) pp.254-255

4.26 M. Suzuki, M. Hirata, T. Itoh: A 165ps/gate 5000-gate ECL gate array. CSSDM (Tokyo) Techn. Dig. (1985) pp.377-380

4.27 Y.H. Chan, J.L. Brown, R.H. Nijhuis, C.R. Rivadeneira, J.R. Struk: A 3ns 32k bipolar RAM. ISSCC Techn. Dig. (1986) pp.210-211

4.28 C.T. Chuang, D.D. Tang, G.P. Li, R.L. Franch, M.B. Ketchen, T.H. Ning, K.H. Brown, C.C. Hu: A sub-nanosecond 5kbit bipolar ECL RAM. Symp. on VLSI Technology, Techn. Dig. (1988) pp.91-92

4.29 F. Buckley, S.Y. Chen, J.K. Hilse, M.E. Homan, G.K. Machol. L. Pereira, J. Terry, G.T. Watanabe: A bipolar 32b processor chip. ISSCC Techn. Dig. (1986) pp.30-31

4.30 P. Thai, S.C. Chang, M.C. Yang: A 35ns 128k fusible bipolar PROM. ISSCC Techn. Dig. (1986) pp.44-45

4.31 D. Chang, C. Schmitz, H. Hingarh, G. Bakker: A 0.9ns ECL 16x4 register file. ISSCC Techn. Dig. (1986) pp.188-189

4.32 E.H. Stevens, W.L. Larson, J.A. Kiddon, B.D. Urke: A bipolar technology for ULSI applications. VLSI Design 6, 92-99 (January 1985)

4.33 A. Matsuzawa, A. Kanda, M. Kagawa, H. Yamada: An 200Msps 8-bit A/D converter with a duplex gray coding. Symp. on VLSI Circuit, Techn. Dig. (1987) pp.109-110

4.34 A bipolar process that's repelling CMOS. Electronics 58, 45-47 (Dec. 23, 1985)

4.35 TI's answer to the need for faster VLSI: its ExCL process. Electronics 60, 73-75 (March 19, 1987)

4.36 K. Sagara, Y. Tamaki, M. Kawamura: Evaluation of dislocation generation in silicon substrates by selective oxidation of U-grooves. J. Electrochem. Soc. 134, 500-502 (1987)

4.37 Y. Tamaki, S. Isomae, K. Sagara, T. Kure, M. Kawamura: Evaluation of dislocation generation in U-groove isolation. J. Electrochem. Soc. 35, 726-730 (1988)

4.38 C.W. Teng, C. Slawinski, W.R. Hunter: Defect generation in trench isolation. IEDM Techn. Dig. (1984) pp.586-589

4.39 K. Hashimoto, Y. Nagakubo, S. Yokogawa, M. Kakumu, M. Kinugawa, K. Sawada, T. Sakurai, M. Isobe, J. Matsunaga, T. Iizuka: Deep trench well isolation for 256kb 6T CMOS static RAM. Symp. on VLSI Technology, Techn. Dig. (1985) pp.94-95

4.40 J. Riseman: Integrated circuit structure with fully enclosed air isolation. U.S. Patent 4106050 (August 1978)

5. A Salicide Base Contact Technology (SCOT) for Use in High Speed Bipolar VLSI

Tadashi Hirao, Tatsuhiko Ikeda, and Yoichi Kuramitsu

LSI Research and Development Laboratory,
Mitsubishi Electric Corporation,
4-1 Mizuhara, Itami, Hyogo, Japan 664

This chapter describes a new process technology called Salicide (self-aligned silicide) base COntact Technology (SCOT), that is applied for realizing high performance prescaler IC and high-gate-density master-slice LSI. The main feature of this process, for reduction of both the base resistance and the capacitance, is the silicidation of the base contact which is opened by employing self-alignment technology. A 1/128, 1/129 two-modulus prescaler IC comprised of 1.5 μm SCOT transistors has been improved to a high operation of 2.1 GHz at 56 mW power dissipation. An ECL 18K-gate masterslice has been developed by a Variable Size Cell (VSC) approach, employing the SCOT process.

5.1 Background

The two-modulus prescaler ICs used for automobile telephones and satellite communication receivers are required not only to operate at a high frequency, i.e., in the GHz band, but also to operate with low power dissipation. The 1.6 GHz 63 mW Si prescaler IC [5.1] and 1.8 GHz 46 mW GaAs prescaler IC [5.2] have been reported. The two-modulus prescaler ICs which are used mainly in digital tuning systems are needed for operation at about 1 GHz, in spite of the fact that each channel has a very narrow band width. The first-stage flip-flops which assume the largest part of the chip power, dissipate about three times more than that of the fixed prescaler IC. In order to enhance the performance of the prescaler ICs, it is therefore important to provide an optimum transistor for high-frequency operation at low power dissipation.

For high-speed data processing systems such as computer mainframes, the ECL masterslice LSIs have been required to increase the integration degree as well as the operating speed. Several types of ECL masterslices [5.3, 4] have been developed in order to meet these requirements. A relatively large number of non-utilized elements, however, remain in these masterslices. Accordingly, it is very important to reduce the non-utilized transistors and resistors for high-gate density and high-operating speed.

The purpose of this chapter is to present a very attractive new salicide base-contact technology (SCOT) process for use in high-speed bipolar VLSI. In the following we describe the gate-speed simulation, the fabrication process and transistor design, the gate-speed and prescaler IC, and the VSC masterslice.

A proposed transistor was realized with the base contact of a self-aligned silicide (salicide) structure, which was made by silicidation of the poly-Si and the silicon surface simultaneously by the same process as MOS technology [5.5]. Using this SCOT process [5.6], the transistor characteristics were improved and consequently a high performance of the two-modulus prescaler ICs was obtained. Furthermore, a VSC approach [5.7] employing the SCOT process was implemented as the most effective process for use in high-speed bipolar VLSI by the ECL 18 K-gate masterslice.

5.2 Gate-Speed Simulation

The principal transistor characteristics limiting the gate speed have been studied by computer-aided simulation [5.8]. The gate speed (propagation delay time t_{pd}) of the ECL inverter was calculated by use of the circuit simulation program SPICE-II [5.9] for the transistor parameters, i.e., the collector base capacitance C_{TC}, forward base transit time τ_F, base series resistance r_B and so on.

An Implanted Self-aligned Contact (ISAC) was chosen as the subject of simulation (Fig. 5.1b) [5.10]. This ISAC process has been specified by the use of a dielectric isolation technology, and a full ion-implantation process with an arsenic buried collector, a boron base and an arsenic implanted self-aligned contact emitter. The process sequence is described as follows: The base is formed by the ion implantation in the transistor region formed by conventional isoplanar technology. Before forming an emitter, all contact windows are opened simultaneously. Then, the base contacts covered with the photoresist mask and the emitters are implanted with arsenic. Annealing the emitter in N_2, the contacts are made into silicide, and the metallization is completed.

80

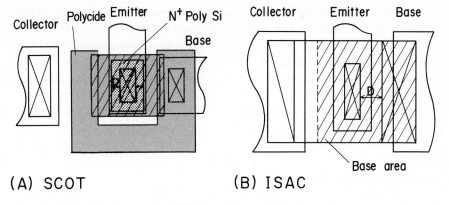

(A) SCOT **(B) ISAC**

Fig.5.1. Schematic layout of a SCOT transistor and ISAC transistor

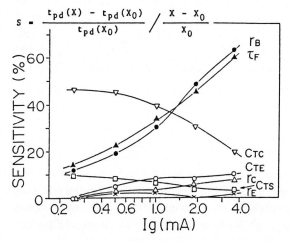

Fig.5.2. Sensitivity of the gate speed as a function of gate current

The initial value X_0 of the parameters were extracted from the ISAC transistor whose emitter size was $2\mu m$ x $4\mu m$. The sequence of simulations $t_{pd}(X)$ was performed by varying the parameters individually. As a result, the contribution of parameters limiting the gate speed was investigated. Figure 5.2 shows the sensitivity S defined as follows:

$$S = \frac{t_{pd}(X) - t_{pd}(X_O)}{t_{pd}(X_O)} \left[\frac{X - X_O}{X_O}\right]^{-1}$$

where X is the relevant parameter of the transistor. The gate current I_g is the sum of the switching and the emitter-follower currents.

81

It is clear from Fig.5.2 that C_{TC}, τ_F and r_B are the main effective transistor characteristics for high speed performance. We should consider such a transistor design because the C_{TC} was reduced at low current operation under 0.3 mA of gate current, and the r_B and τ_F were reduced to 4 mA operation. If the three parameters are reduced at the same time, however, a high-speed gate is realized over a wide current range and also the additive effects of the three parameters are expected at about 1 mA of gate current.

5.3 Process and Transistor Design

Figure 5.1A illustrates the top view of the npn transistor fabricated by the SCOT process (SCOT transistor), and Fig.5.3 shows the key steps in the process sequence of the SCOT transistor.

The isolation of the SCOT transistor consists of the full-recessed oxide and the semi-recessed oxide surrounding the base area. The cross-section of the isolation oxide is a stair structure, as shown in Fig.5.3A. The first p^+ poly-Si layer was formed on the base edge extended over the isolation oxide in order to connect the base Al electrode and silicide contact. The resistor was fabricated at the same time with the p^+ poly-Si layer.

After the base formation, the emitter window was opened and the second poly-Si layer to be deposited over all surfaces was implanted with a heavy dose of As ions, as shown in Fig.5.3B. The emitter region was produced by diffusion from the implanted n^+ poly-Si and was etched off, except the emitter poly-Si electrode. The oxide on the base contact and the p^+ poly-Si were etched away by using the photoresist mask of emitter poly-Si etching, as shown in Fig.5.3C. The thick oxide was selectively grown over the heavily arsenic doped emitter poly-Si in wet oxidation at 820°C by exploiting the phenomenon of concentration-dependent oxidation [5.11]. Therefore, the oxide covering the emitter poly-Si remained after having removed the thin oxide on the base contact and the p^+ poly-Si, and separated the base contact from the n^+ poly-Si.

The Pt-silicide was formed both on the epi-surface and the p^+ poly-Si of base contact, as shown in Fig.5.3D. The base electrode was fabricated with both the silicide of self-aligning opened base contact and the polycide; this can then be called a salicide base contact.

Opening the contact windows of the collector, the emitter at n^+ poly-Si and the base at polycide, and the AlSi metallization completes the processing of the SCOT transistor (Fig.5.3E). The process parame-

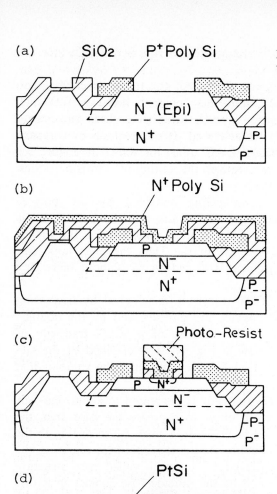

(a) SiO₂ — P⁺ Poly Si

N⁻ (Epi)

N⁺

P⁻

P⁻

(b) N⁺ Poly Si

P

N⁻

N⁺

P⁻

P⁻

(c) Photo-Resist

P — N⁺

N⁻

N⁺

P⁻

P⁻

(d) PtSi

P — N⁺

N⁻

N⁺

P⁻

P⁻

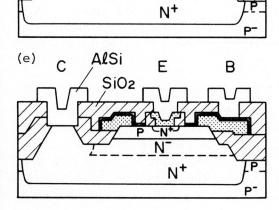

(e) C   AℓSi   E   B

SiO₂

P — N⁺

N⁻

N⁺

P⁻

P⁻

Fig.5.3. Fabrication procedure of the SCOT transistor

ters, the design rule and the dc parameters of the SCOT transistor are listed in Table 5.1

As a result of the gate-speed analysis of the transistor characteristics, the important transistor parameters for high-speed performance are not only the cutoff frequency f_T and the collector-base capacitance C_{TC}, but also the base series resistance r_B. It can be seen by comparing Fig.5.4 with the conventional ISAC transistor shown in Figure 5.1 (B) , that the SCOT transistor realizes the optimum transistor design with these characteristics reduced.

At first, the vertical down-scaling achieved by the poly-Si emitter, in which the emitter depth is 0.1 μm and the base width is 0.13 μm, produces a higher f_T. The shallow junction for the higher f_T, however, causes a rapid increase in r_B. Then the most important problem in improving performance is to reduce r_B without increasing the base area.

The r_B reduction in a SCOT transistor is achieved by decreasing the distance D between base contact and emitter, and by the double-base structure. The distance D, in the case of a SCOT transistor as shown by the cross-sectional view in Fig.5.4, is determined by the salicide base contact structure and is close to 1 μm, half the value of that in an ISAC transistor.

The base area in a SCOT transistor is decreased to about one-half of that in an ISAC transistor. In order to decrease the base area, the emitter length can be decreased to maintain a small r_B according to the double-base structure. The reduction of the parasitic base region in the double-base structure was achieved by the full-walled base structure with the semi-recessed oxide and by the salicide base contact structure. Figure 5.4 illustrates that a SCOT transistor has been synthesized in its design for high performance.

Table 5.1. Device parameters of the SCOT process

Epitaxial thickness	1.6 μm
Base resistance	1 kΩ/square
Base depth	0.23 μm
Emitter depth	0.1 μm
p^+-polysilicon resistance	200, 400, 650 Ω/square
Polycide resistance	8 Ω/square
Contact size	1.5 x 3.0 μm^2
Al line/space	3.5μm/2.0μm
h_{FE}	80
BV_{CEO}	9 V

Synthesized Transistor Design
(Comparison of SCOT and ISAC)

PROCESS PARAMETER	ISAC → SCOT	TRANSISTOR CHARACTERISTICS		
		f_T	C_{TC}	r_B
EMITTER DEPTH	0.4 μm → 0.1 μm	↗	—	↗
EMITTER LENGTH	5.0 μm → 3.0 μm	?	↘↘	↗↗
DISTANCE between BASE and EMITTER	2.0 μm → 1.0 μm	↘	↘	↘↘
BASE STRUCTURE	SINGLE → DOUBLE	?	↗	↘
BASE ELECTRODE	CONVENTIONAL → POLYCIDE	↘	↘↘	—
ISOLATION	LOCOS → STAIRS OXIDE	↘	↘	—
ISAC (IMPLANTED SELF-ALIGN CONTACT)		2.2 GHz	20 fF	92 Ω
SCOT (SALICIDE BASE CONTACT TECHNOLOGY)		4.1 GHz	9 fF	49 Ω

↘↘ : DRASTIC DECREASE   ↘ : DECREASE   ↗ : INCREASE

Schematic Layout of the SCOT Transistor

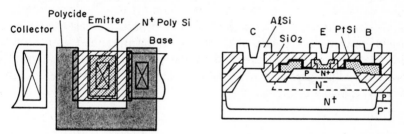

Fig.5.4. Synthesized transistor design by comparison of the SCOT transistor and the ISAC transistor for high performance

The performance of a SCOT transistor used for a prescaler IC was improved, as compared with the current ISAC transistor, that is, the C_{TC} and r_B decreased to half their values and the f_T became twice as high, as can be inferred from Table 5.2. Figure 5.5 shows the f_T of the SCOT transistor with various emitter lengths as a function of the collector current I_c. The maximum f_T obtained was 9.5 GHz for the SCOT transistor in which the emitter consists of four fingers of 1.5 μm x 5 μm.

Table 5.2. Comparison of transistor characteristics for a SCOT prescaler and an ISAC prescaler

	New developments	Current product
Process technology	SCOT	ISAC
Emitter size [μm^2]	1.5 x 3	1.5 x 5
Emitter depth [μm]	0.1	0.4
Capacitance C_{TC} [fF]	9	20
Base resistance r_B [Ω]	49	92
Cutoff frequency f_T [GHz]	4.1	2.2
Delay time of ring-osc. [ps]	140	267
Max. operating frequency [GHz]	2.1	1.1
Power dissipation [mW]	56	125

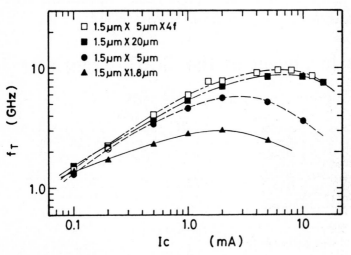

Fig.5.5. Cut-off frequency versus collector current characteristics for various emitter lengths and structures

5.4 Gate Speed and Prescaler IC

Figure 5.6 shows the gate speed t_{pd} of the ECL ring-oscillator employing SCOT transistors with various emitter lengths. When the emitter length was decreased from 20 to 1.8 μm, the t_{pd} became short at low

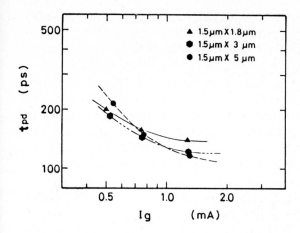

Fig.5.6. Relations between gate speed and gate current in ECL ring-oscillators with various emitter lengths

current because of the decreased C_{TC}, and became long at high current because of increased r_B and decreased f_T. This interpretation is consistent with the result of the simulation shown in Fig.5.2. It is demonstrated in Fig.5.7 that the t_{pd} in the case of a SCOT transistor of the same emitter size ($1.5\mu m \times 5.0\mu m$) is shorter than that of the ISAC one, especially in the high current range, due to the effect of reducing r_B. The minimum t_{pd} achieved was 116 ps at a gate current of $I_g = 1.3$ mA.

According to the curve for the sensitivity in Fig.5.2, each of the improved performances of the SCOT transistor shown in Table 5.2,

Relations Between Gate Speed and Gate Current
in ECL Ring-Oscillators

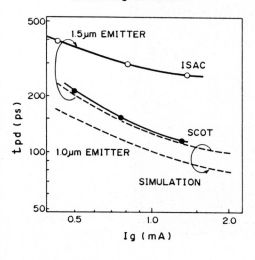

Fig.5.7. Comparison of speed performance of ECL gates employing the SCOT process and ISAC process

leads to a decrease of about 80% in the ECL gate speed. The simulated value $t_{pd}(X)$ of the SCOT circuit became 152ps by using $t_{pd}(X_O) =$ 267 ps for the ISAC circuit, and was in good agreement with the measured 140 ps. Simulating t_{pd} in scaling down the SCOT transistor to 1.0 μm emitter width, high speed can be realized of less than 80ps shown by the dashed line in Fig.5.7.

A 1/128, 1/129 two-modulus prescaler IC was fabricated with 1.5μm SCOT transistors and poly-Si resistors. A microphotograph of the chip is shown in Fig.5.8. The chip size of the SCOT prescaler IC is 1.56mm x 1.56mm. First level metallization of AlSi and polycide cross-under interconnections were employed. Figure 5.9 depicts the relation between input signal level and the operating frequency. The power supply was 5 V and the voltage swing was 280 mV$_{pp}$. This prescaler IC has operated in a wide range from 400MHz to 2.1GHz with 56 mW at an input signal level of -4 dBm.

In Figure 5.10, the performance of a SCOT prescaler IC is compared with other products. The SCOT prescaler IC has four times higher performance than the ISAC prescaler. This prescaler IC operates with half the power dissipation of the reported Si prescaler [5.1] and with the same as GaAs prescaler [5.2]. Decreasing the power dissipation, we obtained 1.4 GHz with 30 mW and 850 MHz with only 19 mW.

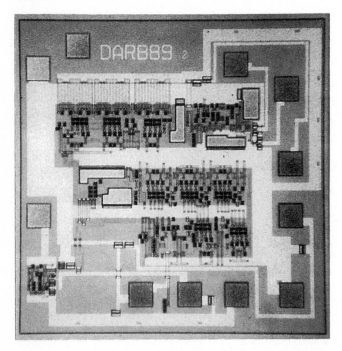

Fig.5.8. Photograph of a two-modulus prescaler

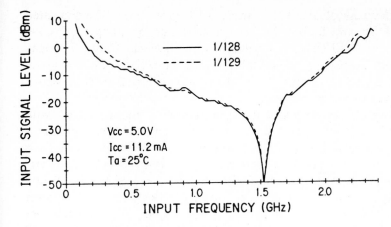

Fig.5.9. Input signal versus operating frequency

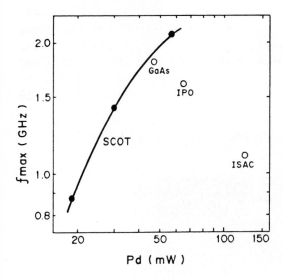

Fig.5.10. Maximum operating frequency as a function of power dissipation

5.5 VSC Masterslice

The actual pattern layout of a two-input OR/NOR gate is shown in Fig.5.11. In the case of the current-cell approach, a relatively large number of the non-utilized elements remain, for example the macro-cell array (MCA) in the simple logic functions. In the VSC (variable size cell) approach, on the other hand, this OR/NOR gate has been achieved without any non-utilized transistor. The full adder with a two-input NOR gate, as an example of complex function, contained only one non-utilized transistor of the 21-component transistors. This

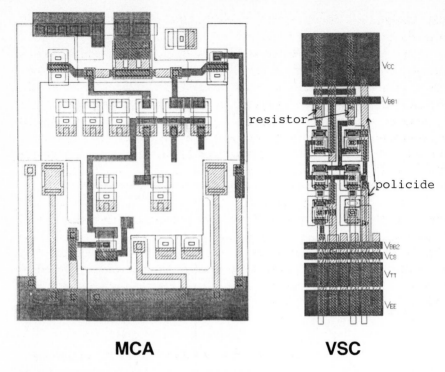

MCA   VSC

Fig.5.11. Layout patterns of two-input OR/NOR gate in macrocell array (MCA) approach and variable size cell (VSC) approach

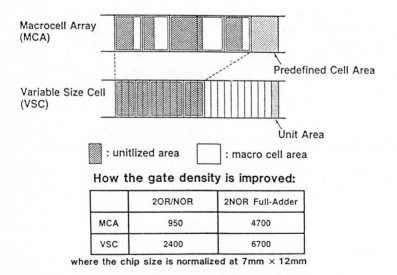

Macrocell Array (MCA)

Predefined Cell Area

Variable Size Cell (VSC)

Unit Area

▨ : unitlized area   ☐ : macro cell area

How the gate density is improved:

	2OR/NOR	2NOR Full-Adder
MCA	950	4700
VSC	2400	6700

where the chip size is normalized at 7mm × 12mm

Fig.5.12. Features of variable size cell (VSC) approach

90

VSC design was found useful to reduce the nonutilized elements in both simple and complex functions.

The VSC concept is based on the design of an array which is constructed from cellular units. This feature of the VSC approach compared with the MCA is illustrated in Fig.5.12. In the VSC approach, the basic units which are composed of three transistors and four poly-Si patterns, are combined with a minimum number to implement each logic function. In order to choose a variable width of each macrocell, not only the circuit wiring but also the resistors are required to determine the proper use during the slice process which customizes a gate-array masterslice.

Figure 5.13 gives a schematic diagram of a SCOT transistor and polycide interconnections used for a VSC masterslice. To realize a VSC structure, the proper use of poly-Si patterns is required during the slice process. The resistor value in each logic cell was determined by the silicidation of a poly-Si pattern during the slice process. In addition, unused poly-Si patterns can be utilized for the polycide interconnection in the intra-cell wiring.

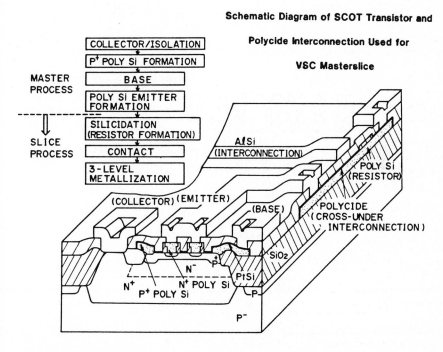

Fig.5.13. Schematic diagram of a SCOT transistor and the polycide interconnection used for the VSC masterslice

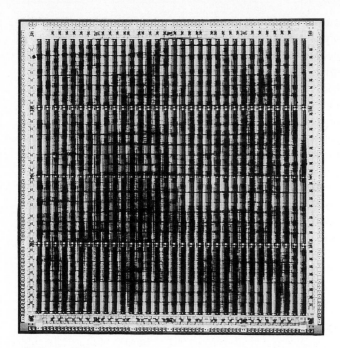

Fig.5.14. Chip photograph of a 32 bit multiplier implemented using the ECL 18 K-gate masterslice

Table 5.3. Synthesized transistor design by comparison of the SCOT transistor and the ISAC transistor for high performance

Technology	1.5 μm rule SCOT
No. of transistors	39,936
No. of poly-Si resistors	53,248
No. of units	13,312
Unit size	24μm x 204μm
Metal pitch	1st 8μm
	2nd 6μm
	3rd 8μm
No. of channels	1,936
No. of I/O pins	256
Interface	ECL 100K compatible
Intrinsic gate delay	150 ps
Supply voltage	V_{EE} = -4.5 V
	V_{TT} = -2.0 V
Switching current	0.4 mA
Emitter-follower current	0.3mA/0.6mA
Chip size	11.90mm x 11.96mm

An ECL 18K-gate masterslice [5.7] was developed by the VSC approach and fabricated by employing a 1.5 μm SCOT process with four-level metallization. The emitter size of the internal transistor was 1.5 μm x 3.0 μm. The first metal layer employing the Pt-polycide was used for the cross-under in a logic cell. The second metal layer of AlSi with a 6 μm pitch wiring and the third layer of AlSiCu were used to interconnect logic cells, and the fourth metal layer of AlSiCu was only utilized for power buses. A chip microphotograph of an ECL 18K-gate masterslice with a 32-bit multiplier implemented is shown in Fig.5.14. The features of the VSC masterslice are summarized in Table 5.3. By implementing the 32-bit multiplier, the gate density was increased by more than 20% in the VSC structure compared with the current-cell structures. The basic gate delay of 150 ps was obtained at a power dissipation of 2.4 mW.

5.6 Conclusion

As an excellent structure with reduced C_{TC}, f_T and r_B simultaneously, the SCOT process employing self-alignment silicide technology has been proposed for use in high-speed bipolar VLSI. A two-modulus prescaler IC consisting of the 1.5 μm SCOT process has achieved a 2.1 GHz operation with 56 mW power dissipation. This result yields higher performance than obtained by GaAs technology to date. A VSC approach which has been realized by a poly-Si resistor and a polycide interconnection of the SCOT process, has increased the gate density by more than 20% compared with the conventional cell structures. An ECL 18K-gate masterslice has been developed and applied to a 32-bit multiplier.

Acknowledgements. The authors would like to thank N. Katoh and T. Nishimura for circuit design, and K. Sakaue and Y. Kinosita for wafer processing. We would also like to thank N. Tsubochi for fruitful discussions and express gratitude to K. Shibayama and H. Nakata for their encouragement.

References

5.1   H. Suzuki, T. Akiyama, K. Ueno: IEDM Techn. Digest (1984) p.682
5.2   K. Maemura, T. Takahashi, S. Inoue, Y. Mitsui, S. Orisaka, O. Ishihara, M. Otsubo: IEDM Techn. Digest (1985) p.94
5.3   W. Brackelmann, H. Fritztsche, H. Ullrich, A. Wieder: IEEE J. SC-20, 1032 (1985)
5.4   M. Tatsuki, S. Kato, M. Okabe, H. Yakushiji, Y. Kuramistu: IEEE. J. SC-21, 234 (1986)

5.5   C.Y. Ting: IEDM Techn. Digest (1984) p.110
5.6   T. Hirao, T. Ikeda, N. Katho: Extended Abstracts of 17th Conf. Solid State
        Devices and Materials (1985) p.381
5.7   T. Nishimura, H. Sato, M. Tatsuki, T. Hirao, Y. Kuramitsu: IEEE J. SC-21,
        727 (1986)
5.8   D.M. Dipietro: ISSCC Techn. papers (1975) p.118
5.9   L.W. Nagel: Electronics Res. Lab. Reports, ERL-M520 (1975)
5.10   Y. Akasaka, Y. Tsukamoto, T. Sakurai, T. Hirao, Y. Horiba, K. Kijima, H.
        Nakata: IEDM Techn. Digest (1978) p.189
5.11   A. Cuthbertson, P. Ashburn: IEDM Techn. Digest (1984) p.749

6. Advanced Self-Alignment Technologies and Resulting Structures of High-Speed Bipolar transistors

Tohru Nakamura, Kazuo Nakazato, Katsuyoshi Washio,
Youich Tamaki, and Mitsuo Namba

ULSI Research Center
Hitachi Central Research Laboratory
Kokubunji, Tokyo 185, Japan

This chapter reports on the development of symmetrical npn transistors with a SIdewall base COntact Structure (SICOS) using advanced self-alignment technology. High cutoff frequencies of 14 GHz in the downward-mode and 4 GHz in the upward-mode operations have been obtained. Using these transistors, high-speed circuits are constructed and their excellent performance is nearly equal to that of GaAs devices. Other high-speed transistor structures are demonstrated, and sub-50ps ECL circuits are predicted.

6.1 Background

Silicon bipolar devices have been used as various integrated circuits such as logic gate arrays, analog LSIs, and static memory LSIs. The advantage of bipolar devices is their high-speed performance. Conventional device structures, which include isoplanar or U-grooved isolation type transistors, have large parasitic areas that affect high-speed operation. In order to reduce device size and the accompanying parasitic components, self-aligned device structures have been developed [6.1-6]. The trend in ECL circuit gate delays is shown by the curve in Fig.6.1. The delay time of ECL circuits constructed with conventional transistors was approximately 200 ps, and it was difficult to obtain gate delays below 100 ps. The polysilicon self-aligned bipolar device structure, which includes polysilicon base transistors like SST [6.5] or SICOS [6.1], have been developed to attain high-speed performance which eliminates parasitic areas. Since 1984 when a sub-100ps gate delay was obtained by ECL circuits [6.3], new self-aligned device structures having high-speed performance below 50 ps have been proposed [6.7,8]

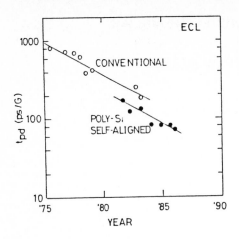

Fig.6.1. Evolution of ECL circuit gate delay

Polysilicon base transistors [6.2, 6] are fabricated by various self-aligned process technologies to eliminate the parasitic areas. In most of these transistors, emitter and intrinsic base areas are defined by self-aligned technologies using polysilicon electrodes, but total base areas, including extrinsic base regions, are determined by other mask dimensions. Super Self-aligned process Technology (SST) [6.5] and SIdewall base COntact Structure (SICOS) [6.1, 3, 4] transistors are also self-aligned bipolar devices using polysilicon base electrodes. In SST transistors, the total base and emitter areas are defined by one mask dimension. In SICOS transistors, total base and emitter areas, as well as isolation areas, are determined by only one mask dimension, too. Moreover, the symmetric structure of a bipolar transistor is suitable for reducing parasitic areas. The SICOS transistor is one of the most promising candidates for an ideal transistor.

In this chapter, the advanced self-alignment techniques used to fabricate SICOS transistors and the high-speed characteristics of the transistors are demonstrated. Prospective characteristics of high-speed silicon bipolar transistors structures are also discussed.

6.2 SICOS Device Structure

Recent bipolar transistor structures are illustrated in Fig.6.2. With the conventional structures shown in Fig.6.2a, bipolar operations are basically limited to the intrinsic regions, while the extrinsic base regions are used as contact areas for electrodes. Usually the areas of the extrinsic regions are much larger than those of the intrinsic ones. Therefore, parasitic factors in the extrinsic regions are likely to obscure high performance in the intrinsic regions. Especially critical are

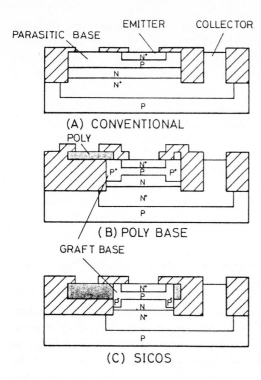

PARASITIC BASE   EMITTER   COLLECTOR

Fig.6.2. Recent bipolar transistor structures

(A) CONVENTIONAL

POLY

(B) POLY BASE

GRAFT BASE

(C) SICOS

the large capacitances between base and collector regions for npn transistors. In a conventional structure fabricated with 3 μm design rules, the areas occupied by the intrinsic base regions are approximately 1/3 to 1/10 of the sizes utilized for the extrinsic base regions. Parasitic capacitance between base and collector influences the cutoff frequency in a npn transistor. This indicates that high-speed performance is attainable if the base regions can be considerably reduced.

To improve high-speed characteristics, polysilicon base transistor structures have been developed. To reduce the extrinsic regions or parasitic areas, polysilicon base electrodes, contacted on the graft base areas surrounding the emitter area, were used on the thick SiO_2 layer, as shown in Fig.6.2b. In the polysilicon base transistor structures, the capacitances between base and collection regions are much lower than those of conventional structures. In the structure under consideration, which uses self-aligned technology, polysilicon base electrodes and intrinsic base and emitter regions are determined by only one photomask pattern. Therefore, high-speed operations have been achieved through the reduction of parasitic base capacitance. But deep graft base areas connected to both polysilicon electrodes and intrinsic base regions remain parasitic base areas.

To obtain higher-speed characteristics by reducing the parasitic areas, the elimination of parasitic-base, or graft-base regions is needed. Moreover, as is described in Sect.6.5, reduced parasitic base areas and contact areas of the polysilicon base electrodes to intrinsic base regions are highly influential with regard to high-speed operation. In the SICOS transistor structure, as shown in Fig.6.2c, only the active transistor region is formed in the single crystal silicon epitaxial layer. Base electrodes, which are connected to the sidewall of the intrinsic base, are made by laying polysilicon over the thick SiO_2 layers. SICOS transistors are fabricated by advanced self-aligned techniques. These techniques enable one to use a single photomask pattern to make the intrinsic and extrinsic base regions, emitter region, base and emitter contact areas, and isolation areas between adjacent transistors. The parasitic areas are reduced as much as possible to eliminate the parasitic diode effect, resulting in a higher cutoff frequency. The SICOS devices feature symmetric structures enabling bipolar transistors to exhibit ideal, one-dimensional, operations.

6.3 Fabrication Process

The process steps necessary to fabricate SICOS device structures are shown in Fig.6.3.

An n-type epitaxial layer of approximately 1 Ω·cm is grown on the Sb-doped buried layers over the p-type substrate. SiO_2, Si_3N_4, and SiO_2 layers are successively formed on the surface of the epitaxial layer. Emitter, collector, and resistor regions are defined by photoresist patterns. Device areas and isolation areas are determined by these patterns. The tri-level layers of SiO_2, Si_3N_4, and SiO_2 and the silicon epitaxial layer under the photoresist patterns are removed by reactive sputter-etching, as shown in Fig.6.3a

Thermal oxidation and Si_3N_4 deposition are then performed on all surfaces. This is followed by reactive ion etching to remove emitter and collector plateau regions except at the plateau region sidewalls. Selective oxidation leads to the formation of thick-field SiO_2 layers, as shown in Fig.6.3b. The silicon dry-etched area remaining above the thick-field oxide becomes the base contact regions which are connected to the polysilicon electrodes. The shape of the plateau regions is important because the generation of crystal defects is highly dependent on the shape of the silicon surface. Smooth silicon surfaces created by dry etching and optimum thickness of the oxide layers inhibit crystal defects.

After the Si_3N_4 and thin SiO_2 layers at the sidewalls have been removed, an undoped polysilicon film is deposited. The polysilicon

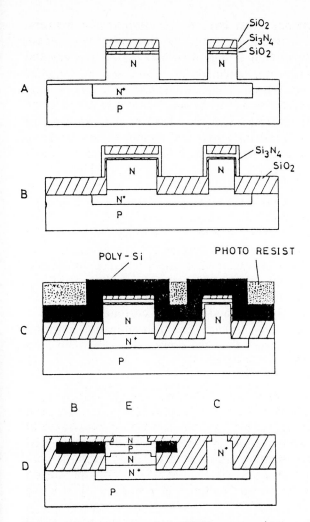

Fig.6.3. Fabrication steps for a SICOS npn transistor

film over the plateau regions must be completely eliminated to fabricate polysilicon electrodes. The self-aligned photoresist etching-back process is used to construct the electrodes.

Photoresist films are coated onto the polysilicon layer to obtain flat photoresist surfaces. This is followed by O_2 plasma sputter-etching to expose the polysilicon layer on the plateau regions. Exposed polysilicon regions are then removed by sputter etching using a SF_6+H_2 gas mixture, as shown in Fig.6.3c.

Boron ions are then implanted into the polysilicon layers, followed by selective oxidation to isolate the base electrodes from the emitter electrodes and to diffuse the graft base region. Boron ions are implanted into the epitaxial layers to form the intrinsic base region. Finally, a Si_3N_4 passivation film is deposited and the emitter window opened. The emitter regions are defined by arsenic ion implantation through the thin polysilicon layer. Contact windows are then opened and aluminum electrodes are formed, as shown in Fig.6.3d.

Advanced self-alignment techniques are used to decrease the size of the parasitic regions. The intrinsic base and emitter regions and isolation regions are defined using the same photomask pattern.

6.4 Electrical Characteristics

6.4.1 Transistor

Gummel plots of the SICOS transistor operated in downward and upward modes are shown in Fig.6.4, where the reverse bias between collector and base is 0 V. The base and collector currents decreased to less than 10^{-10} A without any leakage and recombination currents. Upward current gains were larger than the downward ones when mea-

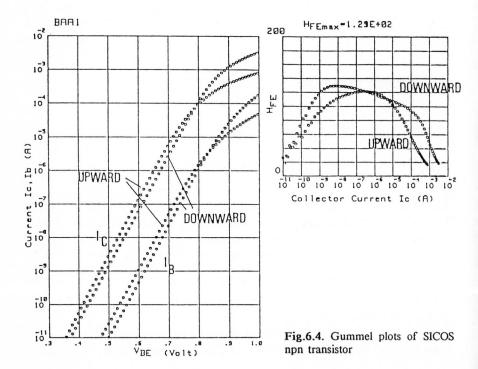

Fig.6.4. Gummel plots of SICOS npn transistor

sured at the low current level. This means that SICOS transistors have a symmetric structure, and the Gummel number in the surface n-type region (emitter regions in the downward-operated transistors) is larger than that in buried layers (emitter regions in the upward-operated transistors). Also, the base Gummel number of upward operated transistors decreases, because the base profile is not symmetric and a depletion layer of the base regions from the collector-base junction in an upward-operated transistor is smaller than that in a downward-operated one.

The cutoff frequencies as a function of the collector currents for SICOS npn transistors are shown in Fig.6.5. The emitter area is 6 μm^2, and an epitaxial layer of 0.7 μm was adopted. The maximum cutoff frequency is 14 GHz for downward operated transistors, and 4GHz for upward ones. Conventional device structures generally have very different characteristics for downward and upward operation. Upward-operated transistors have lower characteristics than the downward-operated ones. This is because the junction area between emitter and base is much larger than that between base and collector. SICOS transistors have very high upward cutoff frequencies, that are slightly lower than the downward cutoff frequencies. High upward cutoff frequency transistors are applied to IIL circuits and those with high downward cutoff frequencies are used with high-speed ECL circuits.

Experimental results for SICOS npn transistors are compared with those of conventional devices in Table 6.1. SICOS transistor areas are reduced by less than 1/3 for the same emitter dimensions as conven-

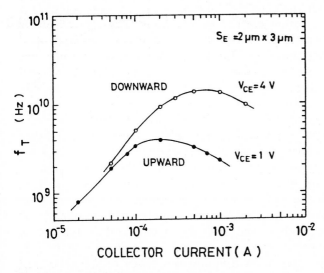

Fig.6.5. Cutoff frequency as a function of npn transistor

Table 6.1. SICOS npn transistor characteristics

	SICOS	Conventional
Emitter area	$2 \times 3 \ \mu m^2$	$2 \times 3 \ \mu m^2$
Cell area	$128 \ \mu m^2$	$460 \ \mu m^2$
h_{FE} (downward)	50	80
h_{FE} (upward)	70	5
C_{CB}	8 fF	25 fF
C_{EB}	18 fF	27 fF
C_{CD}	15 fF	45 fF
$r_{bb'}$	500 Ω	700 Ω
f_T (downward)	14 GHz	7 GHz
f_T (upward)	4 GHz	0.05 GHz

tional transistors. This is due to the self-alignment techniques which reduce mask tolerance between isolation and base regions. Self-alignment techniques lower the parasitic capacitance, such as the capacitance between collector and base region C_{CB}, to below that of conventional transistors.

6.4.2 Circuits

Measured IIL and ECL gate delays fabricated on the same chip as a function of power dissipation are shown in Fig.6.6. The results are for FI = FO = 1. The emitter dimensions are 6 μm^2. A minimum gate delay of 84 ps/gate was measured for ECL circuits at a collector current of 0.9 mA. For the IIL circuits, the minimum gate dealy was 320 ps/gate at a power dissipation of 5 mW. A gate delay of 600 ps/gate was measured for IIL circuits at 50 μW/gate. This indicated that low-power VLSIs could be constructed with SICOS IIL circuits.

High-speed SICOS transistors have been applied to frequency divider circuits. Frequency divider circuits constructed by ECL gates contain 4-stage divide-by-two circuits, three internal buffers, and an output buffer, as shown in Fig.6.7. Each D-type master-slave flip-flop was constructed using the ECL series gate technique. The current level of each transistor was 1 mA, at which maximum cutoff frequency was achieved. The supply voltage was 5 V. Input and output waveforms and the relation between input power and dividing frequency are shown in Fig.6.8. The maximum dividing frequency of 6.9 GHz was attained when the input power was 10 dBm.

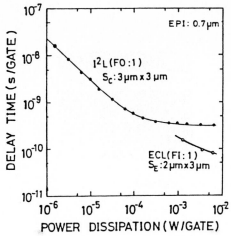

Fig.6.6. Delay time of ECL and IIL circuits as a function of power dissipation

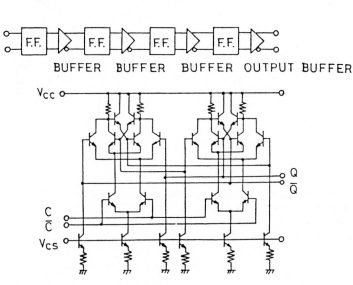

ECL F.F.

Fig.6.7. ECL frequency divider circuits

IIL frequency divider circuits were also fabricated. Each divide-by-two circuit contains seven IIL gates. IIL circuits operated at more than about 0.1 mW/gate consume more power, as shown in Fig.6.6, because of degraded current gains in lateral pnp transistors. A maximum dividing frequency of 580MHz was obtained at a power/stage of 9 mW, as shown in Fig.6.9. The features of frequency divider circuits

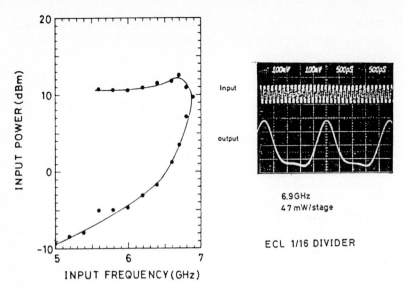

Fig.6.8. Input characteristics of high-speed ECL frequency divider circuits

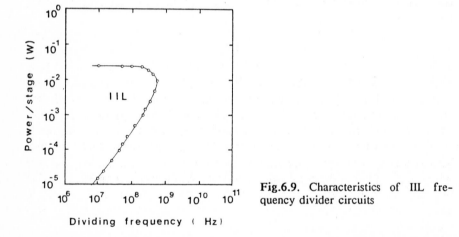

Fig.6.9. Characteristics of IIL frequency divider circuits

fabricated by SICOS transistors are low power dissipation and high speed. This high-speed performance is close to that of GaAs devices.

6.5 Advanced Process Technology and Electrical Results for High-Speed SICOS Transistors

SICOS transistors have a nearly symmetric structure indicating that the emitter areas are almost equal to the collector areas. However, small

104

parasitic areas still exist in present devices. The remaining parasitic areas are graft base regions, where the intrinsic base is connected to the polysilicon electrodes. It is necessary to fabricate these regions to reduce base series resistance r_{bb}'. Graft base areas affect breakdown characteristics and junction capacitances between base and collector regions. BV_{cbo} of SICOS transistors is smaller than that of conventional transistors. To fabricate actual ideal bipolar devices, graft base areas must be reduced as much as possible. In order to obtain higher-speed ECL circuits, the delay time components in SICOS transistors have been analyzed.

The simulated ECL gate delay time components of SICOS transistors operating at maximum speed are shown in Fig.6.10. In the conventional structure, a minimum gate delay of 170 ps/G was governed by the junction capacitance and an intrinsic transistor component characterized by cutoff frequency and base resistance. Approximately half of one gate delay was determined by the junction capacitance. In the previous SICOS transistor each component was significantly reduced, but the percentage contribution of each component to the total gate delay was almost the same as that of the conventional structure. To obtain higher-speed ECL circuits, junction capacitances must be reduced.

The schematic of the electron flow in a SICOS transistor is depicted in Fig.6.11. At a low current level, electrons injected from the emitter to the base region flow uniformly through almost all base regions. Since the base resistance is neglected at this level, the collector current is almost proportional to the base areas. When the collector current increases, electron flow in the inner base regions is reduced. Meanwhile, since outer base regions are constructed with a deep graft base area, almost all electrons flow through the thick base regions. Therefore, the cutoff frequency and current gain gradually decrease when the collector current increases. In self-aligned transistors using polysilicon base electrodes, graft base regions are more sensitive to the ac and dc characteristics compared with those of the conventional transistors.

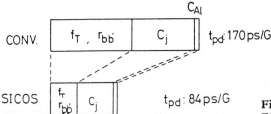

SIMULATED ECL GATE DELAY TIME

Fig.6.10. Simulated gate delay of ECL circuits consisting of conventional and SICOS transistors

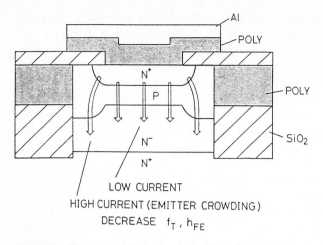

Fig.6.11. Electron path in the polysilicon base transistors at low and high current levels

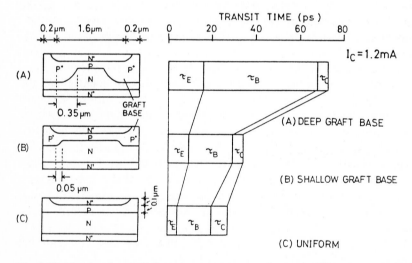

Fig.6.12. Simulated cutoff frequency components in three different transistors

The influence of the graft base regions on the cutoff frequencies at a high current level were simulated. The cutoff-frequency components of three npn transistors are shown in Fig.6.12. Reciprocal cutoff frequency is represented by the transistor's transit time. The simulations were conducted using a collector current of 1.2mA and an emitter area of 1.6μm x 4μm. Cross-sectional views of transistors are also drawn. A transistor having deep graft base regions is illustrate in Fig.6.12a. Graft base regions are formed under the emitter region. The overlap length is 0.35 μm from the emitter mask. A transistor with shallow graft base regions is shown in Fig.6.12b. The overlap length here is only 0.05 μm. A transistor without graft base regions is ehibited in Fig.6.12c.

The transit time is determined [6.9] by

$$\tau = \tau_E + \tau_B + \tau_C ,$$

where T_E is the emitter time delay, τ_B the base time delay, and τ_C the collector time delay. The base transit time delay of transistors was found to be large for transistors with deep graft base areas, and was dramatically diminished by reducing graft base regions from 0.3 to 0.5 μm. The smallest total transit time was obtained for the transistor without graft base regions, but it is difficult to fabricate an actual transistor without graft base regions. This indicates that graft base regions should be made as shallow as possible to obtain high-speed bipolar transistors.

The fabrication of fine graft base regions for higher-speed advanced SICOS transistors requires new processes. Two-step sidewall oxidation techniques to make fine base contact regions and shallow graft base regions were developed, as shown in Fig.6.13. After the tri-level layers and the silicon epitaxial layer were removed by reactive sputter-etching, the first sidewall thermal oxidation with a thickness of

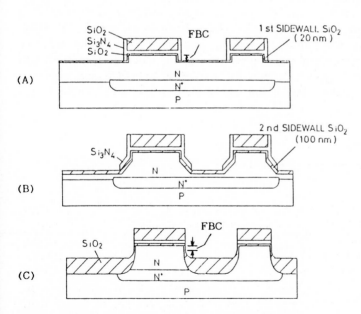

Fig.6.13. Fabrication steps of advanced SICOS transistors

30 nm was carried out, and Si_3N_4 layers were formed over the silicon surface. The Si_3N_4 layers were removed at the sidewalls by reactive etching, as shown in Fig.6.13a. Next, the second 100 nm thick sidewall thermal oxidation and Si_3N_4 re-deposition were performed, followed by reactive ion etching to remove Si_3N_4 at the sidewalls, indicated in Fig.6.13b. Selective oxidation followed to form thick field SiO_2 layers. The first thin oxidized regions became polysilicon base contacts to the intrinsic base which are denoted by "FBC" in Fig.6.13. After these steps, conventional SICOS fabrication processes were used to make advanced SICOS transistors with shallow graft base regions, as described above and in Fig.6.3.

The characteristics of advanced SICOS npn transistors are summarized in Table 6.2. A fine photolithography technique was used to make small emitter dimensions. The transistor area inside the isolation area was 85 μm^2 with an emitter mask dimension of $1.5\mu m$ x $4\mu m$. A small junction capacitance of 5fF between the collector and base regions was attained by greatly reducing the graft base regions. A high cutoff frequency of 12 GHz was obtained at a collector and emitter voltage of 1 V, and a frequency of 16 GHz was obtained at 3 V. A high breakdown voltage of 11 V between collector and base electrodes was also achieved. These results were derived from reduced graft base structures of advanced SICOS transistors.

Photographs of 25-stage ECL ring oscillators made from advanced SICOS transistors are shown in Fig.6.14. The emitter size in the ECL circuits is $0.8\mu m$ x $3.3\mu m$. ECL gates were designed to operate at a V_{CC} of 2V and a V_{EE} of -1.5 V. The voltage swing was 0.5 V. Polysilicon resistors were used to reduce gate delays caused by parasitic capacitance. The output waveforms are shown, too. The minimum gate delays obtained were 63 ps/gate for a FI = 1. The ECL gate delays as a function of power dissipation are shown in Fig.6.15. These results

Table 6.2. Advanced SICOS npn transistor characteristics

Emitter area	0.8×3.3 μm^2
Cell area	85 μm^2
C_{CB}	5 fF
C_{EB}	9.6 fF
C_{CS}	18 fF
f_T ($V_{CE}=3V$)	16 GHz
BV_{CEO}	5.5 V
BV_{CBO}	11 V
BV_{EBO}	4 V
EPI	0.8 μm, 0.5 $\Omega \cdot cm$

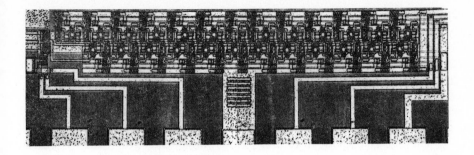

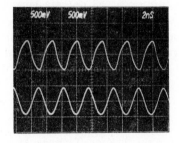

ECL

FI = 1

t_{pd}: 63 ps/G

$A_E = 0.8 \times 3.3\ \mu m^2$

Fig.6.14. Schematics of advanced SICOS 25-stage ring oscillators

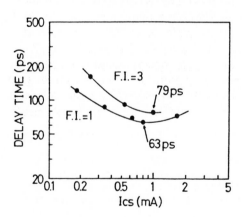

Fig.6.15. Delay time of ECL circuits as a function of switching n current

were measured for FI = 1 and FI = 3. A minimum gate delay of 63 ps/gate was measured for FI = 1 at a collector current of 0.8 mA, and 79 ps/gate for FI = 3 and 0.95 mA.

6.6 Conclusions

The characteristics and features of SICOS transistors have briefly been described. SICOS transistors feature symmetric structures and reduced parasitic capacitance, which allow high-speed performance. Advanced self-aligning techniques to reduce the parasitic regions have been developed, and ECL circuit speeds far below 100 ps/gate have been obtained.

Advanced SICOS structures which have fine graft base regions enable high-speed characteristics in self-aligned bipolar devices. Two-step oxidation technologies have been developed to make accurate, shallow graft base regions. ECL circuits of 63 ps/gate have been obtained in the advanced SICOS structures. Future devices, including SICOS transistors, will have reduced graft base regions, shallow emitter and base junctions, and will become nearly ideal, one-dimensional structures.

References

6.1 T. Nakamura, T. Miyazaki, S. Takahashi, T. Kure, T. Okabe, M. Nagata: Self-aligned transistor with sidewall base electrode. ISSCC Dig. Tech. Papers (1981) pp.214-215

6.2 T.H. Ning, R.D. Isaac, P.M. Solomon, D.D. Tang, H.N. Yu: Self-aligned npn bipolar transistors. IEDM Tech. Dig. (1980) pp.823-824

6.3 T. Nakamura, K. Nakazato, T. Miyazaki, T. Okabe, M. Nagata: Integrated 84 ps ECL with IIL. ISSCC Dig. Tech. Papers (1984) pp.152-153

6.4 T. Nakamura, K. Ikeda, K. Nakazato, K. Washio, M. Namba, T. Hayashida: 63 ps ECL circuits using advanced SICOS technology. IEDM Tech. Dig. (1986) pp.472-475

6.5 T. Sakai, Y. Kobayashi, H. Yamauchi, M. Sato, T. Makino: High speed bipolar ICs using super self-aligned process technology. Proc. 12th Conf. on Solid State Devices (Tokyo, 1980) pp.67-68

6.6 T. Sakai, S. Konaka, Y. Yamamoto, M. Suzuki: Prospects of SST technology for high speed LSI. IEDM Tech. Dig. (1985) pp.18-21

6.7 K. Washio, T. Nakamura, K. Nakazato, T. Hayashida: A 48 ps ECL in a self-aligned bipolar technology. ISSCC Dig. Tech. Papers (1987) pp.58-59

6.8 H.K. Park, K. Boyer, C. Clawson, G. Eiden, A. Tang, T. Yamaguchi, J. Sach-itano: High speed polysilicon emitter-base bipolar transistors. Electron. Dev. Lett. EDL-7, 658-660 (1986)

6.9 R.L. Kronquist, J.Y. Fourrier, J.P. Pestie, M.E. Brilman: Determination of a microwave transistor model based on an experimental study of its internal structure. Solid-State Electron. 18, 949-963 (1975)

7. Trends in Heterojunction Silicon Bipolar Transistors

M. Ghannam[1], J. Nijs, and R. Mertens

Interuniversity Microelectronics Center IMEC, Kapeldreef 75
B-3030 Leuven, Belgium

Ultrafast bipolar transistors have an extremely thin base and therefore should follow a vertical scaling scheme. The vertical scaling of silicon bipolar transistors, however, necessitates the use of novel emitter structures. Self-aligned polysilicon emitter and base contact transistors suffer from two basic problems: 1) the necessary high-temperature treatment to reduce the emitter series resistance resulting in a smaller emitter efficiency and outdiffusion of the polysilicon dopant into the underlying monocrystalline silicon, and 2) two-dimensional issues related to base encroachment problems resulting in a trade-off between emitter-base tunneling, emitter-collector punchthrough and transient delay. Epitaxial emitters can solve some of these problems provided the emitter is grown at low temperatures. New results concerning low-temperature epi-growth of emitters are presented. The advantages of heterojunction emitter-base junctions are summarized. Results and shortcomings of wide-bandgap emitters are discussed. Narrow-bandgap base heterojunction silicon transistors are shown to be a very interesting novel approach.

7.1 Background

As a result of the enormous improvement achieved in microelectronics technology during the last twenty years, it has become possible to combine Very Large Scale Integration (VLSI) and very high speed of operation. The minimum lateral dimension has been scaled down from $25\mu m$ in 1960 to $1\mu m$ in 1986. In bipolar transistors, lateral scaling reduces parasitics caused by the lateral extension of the device but does not significantly affect the intrinsic device. To obtain higher-speed performance, the vertical dimensions of the device must also be reduced. Modern high-speed bipolar transistors therefore have an extremely shallow emitter and a very thin base. Very shallow emitters

---

[1] Also with the Electronics and Communications Department, Cairo University, Guiza, Egypt.

are accompanied by strongly reduced emitter efficiency. To avoid this problem, research has been oriented towards non-conventional metal contacts. The polysilicon emitter contact has been proposed as a new emitter contact scheme that maintains a high emitter efficiency with very shallow structures. Collector to emitter punchthrough and spatial base dopant fluctuations are major problems associated with the very thin base region. Doping the base heavily reduces the importance of these two problems but induces tunneling currents at the emitter base junction [7.1] and increases the emitter base junction capacitance. As we shall see later, the tunneling-punchthrough trade-off can be removed in heterojunction bipolar transistors. Keeping the bulk and contact resistances at very small values becomes a stringent requirement of scaled VLSI bipolar transistors. As the device area becomes smaller and smaller, the operating current density might reach 10^5 A/cm^2 fixing the upper limit of the emitter specific resistance at $10^{-7}\,\Omega\cdot$cm^2 [7.2]. Device physics related to surface geometry and to lateral effects are among the factors that must be considered in scaled transistors. For example, in advanced self-aligned sidewall spacer bipolar technology, the effects of the extrinsic-intrinsic base encroachment [7.3] on current gain, punchthrough, tunneling and transient delay become first-order problems. Thus, two-dimensional analysis becomes necessary. By using elaborate emitter schemes bidimensional effects can also be minimized

In short, a perfect high-speed, scaled bipolar transistor cannot be realized by conventional bipolar technology due to the many undesirable existing trade-offs. A key parameter in relaxing these trade-offs is the transistor's emitter efficiency. The design and realization of a high emitter efficiency bipolar transistor using new structures, materials or concepts have always been, and will continue to be, a challenge. In this chapter, the properties of different types of novel emitter structures will be reviewed. Section 7.2 deals with polysilicon emitters whereas epitaxial emitters will be discussed in Sect.7.3. Finally the actual status of heterojunction silicon bipolar transistors will be described in Sect.7.4 and important areas for future research will be identified.

7.2 Polysilicon Emitter Bipolar Transistors

The first experiments concerning polysilicon emitters for npn transistors aimed at the use of this material, doped during deposition in the DOPOS technique [7.4] as a diffusion source for very shallow emitter structures. Later on, the POLYSIL technique [7.5] was used for the deposition of an undoped polysilicon film, afterwards heavily doped by ion implantation. Subsequent high-temperature annealing results in a net reduction of the film resistance and in the formation of a shallow emitter in the crystalline region. The final structure is represented schematically in Fig.7.1. The current gain of the transistors made by this technique is improved by one order of magnitude over the conventional value. The reason for this improvement was first attributed to a reduced bandgap narrowing in polysilicon [7.5]. This explanation is not very plausible, however, since bandgap narrowing occurs inside the doped crystallites. Another reason for the improved injection efficiency might be the presence of a thin interfacial oxide between the polysilicon film and the substrate [7.6]. This oxide creates a barrier for hole injection into the emitter by reducing the hole tunneling probability. It also creates a band bending that favors electron injection from emitter to base and impedes hole injection from base to emitter. Finally, it has been proposed that the smaller hole mobility in n-type doped polysilicon is responsible for the reduced hole injection [7.7]. A unified theoretical model for the current transport in polysilicon [7.8] showed that the transport in the polysilicon layer might play a role in determining hole injection only at interfacial oxide thickness less than 15 Å. On the other hand, the hole current is totally controlled by the tunneling probability if the interfacial oxide is thicker than 20 Å. This conclusion has been confirmed experimentally by exploring the effect of surface treatment prior to polysilicon deposition on the current gain of the transistor [7.9]. The transistors given an RCA clean (with chemical oxide grown at the interface) had a current gain approximately five times higher than those given an HF etch just prior to polysilicon deposition.

Polysilicon emitters are widely used in state-of-the-art silicon bipolar transistor technology. Besides improved emitter efficiency, they offer the possibility of self-aligned processes strongly required for scaling purposes [7.10]. The emitter resistance is, however, a major drawback associated with this type of transistor. Recent measurements

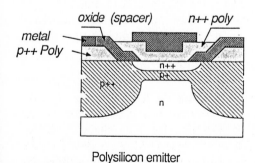

Polysilicon emitter

Fig.7.1. Schematic representation of an advanced self-aligned polysilicon-emitter bipolar transistor which uses an SiO$_2$ sidewall spacer between the emitter contact and the base contact

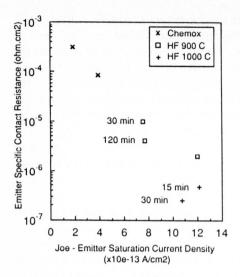

Fig.7.2. Correlation between the specific resistance of an n⁺ polysilicon emitter contact and the hole emitter saturation current density J_{oe}. These results are determined experimentally by *Crabbé* et al. [7.11]

of the specific emitter contact resistance in polysilicon emitters revealed an inverse dependence on the base component of the emitter reverse saturation current (J_{oe}) [7.11], as shown in Fig.7.2. Small contact resistances are obtained only with no intentionally grown interfacial oxides. They are associated with reduced emitter efficiency and can only be obtained after relatively long anneals at high temperatures. These heat treatments result in deeper crystalline emitters and eventually in epitaxial alignment of most of the polysilicon film [7.12, 13]; the advantage of using a polysilicon emitter contact becomes seriously questionable. Besides significant resistance, polysilicon emitters suffer from yield problems caused by the interfacial oxide and by epitaxial alignment. Also, the emitter delay can be longer than for a conventional metal contact (due to the rectangular minority carrier distribution in the crystalline emitter region) and contributes to a significant portion of the total delay of the transistor. In addition, during high temperature annealing, lateral diffusion of the emitter might occur and hence the problems related to base encroachment are not removed. Moreover, since the emitter is heavily doped, the trade-off tunneling-punchthrough is not eliminated in scaled polysilicon-emitter bipolar transistors.

7.3 Epitaxial Emitter Bipolar Transistors

Crystalline emitters can be grown epitaxially. In a non-selective epitaxial emitter growth an n-type silicon epitaxial layer is deposited in windows opened in the oxide. Silicon deposited elsewhere is polycrystalline and is etched after defining the emitter edges. Atmospheric pressure chemical vapor deposition epitaxy (APCVD) can be used. It is, however, processed at a high temperature (>1050°C) and therefore suffers from autodoping and dopant outdiffusion into and out of the substrate. Outdiffusion commonly occurs with this technique and causes a yield problem, especially with the very thin base regions involved. Also, the emitter becomes doped by compensation near the junction, which degrades the material quality, and hence the carrier lifetime. Moreover, lateral outdiffusion also occurs, which results in a lateral extension of the emitter region, as shown in Fig.7.3. Hence, like the polysilicon emitter, APCVD epitaxial emitters do not eliminate bidimensional problems.

Epitaxial emitters can also be made by aligning polysilicon deposited at a temperature around 600°C epitaxially to the crystalline substrate [7.12, 13]. This alignment occurs after a post-deposition high-temperature treatment at 800°-900°C which is a lower temperature than for APCVD epitaxy. In this case the emitter is doped by implantation and dopants outdiffuse into the base during the alignment and annealing heat cycles. In comparison with APCVD epitaxial emitters, the quality of the aligned polysilicon material is degraded.

Low-temperature epitaxy is a very atractive approach for obtaining a well-controlled and compatible emitter technology. Different processing techniques have been suggested and can be classified into three categories depending on the deposition temperature. The first method uses plasma-enhanced chemical vapor deposition at 750°C for the deposition of an undoped epitaxial film [7.14]. Doping of the films is obtained by subsequent ion implantation. A high-temperature step (at 950°C) is required to activate the implanted ions. This step should be as short as possible in order to accurate control the very shallow doping profiles.

Epitaxial deposition of boron-doped Si films at 550°C [7.15] and of undoped Si films at T < 750° [7.16] by ultrahigh vacuum chemical

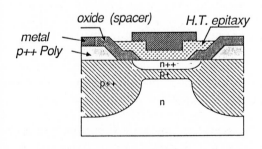

High temperature epitaxial emitter

Fig.7.3. Schematic representation of an advanced self-aligned APCVD epitaxial-emitter bipolar transistor which uses an SiO_2 sidewall spacer between the emitter contact and the base contact

vapor deposition has been reported. However, as in all Low Pressure Chemical Vapor Deposition (LPCVD) processes, doping the film uniformly with phosphorus (in situ) can be a problem.

Finally, deposition of low-temperature epitaxial layers at 250°-300°C is feasible by glow-discharge decomposition of silane (SiH_4) [7.17]. The layers can be doped in-situ by introducing the gas containing the dopant (PH_3 for phosphorus doping) into the gas mixture. In Fig.7.4a, a cross-sectional TEM photograph shows that the grown film is partially epitaxial and partially amorphous (a-Si:H). The boundary between the crystalline region and the amorphous region is very clear. In some localized areas the deposited film is totally crystalline over all its thickness (50nm), as shown in Fig.7.4b. After a 700°C anneal for 20 min, the film totally aligns epitaxially and has a very respectable crystalline quality with very few defects (twins), as can be seen in Fig.7.4c. The number of defects (twins) in the regrown a-Si:H is strongly correlated with the amount of oxide present at the a-Si:H/crystalline-silicon original interface. The presence of twin defects does not strongly degrade the electrical quality of the epitaxial material because these defects do not constitute effective centers for carrier recombination [7.18]. Since the processing temperature is limited to 700°C, the transition from the n-type emitter profile to the p-type base profile is very abrupt. Other low-temperature epitaxial techniques include molecular beam epitaxy [7.19], LPCVD [7.20] and photo-enhanced epitaxy [7.21].

Among the advantages offered by low-temperature epitaxial emitters we enumerate: 1) small emitter bulk and contact resistances; 2) in situ doping results in very abrupt profile transition from emitter to base; an accurate control of the base width and consequently an excellent reproducibility and a high yield are expected; also the retardation field slowing down electron motion near the emitter base junction is eliminated; 3) high-quality emitters due to the non-compensated nature of the material; 4) suppression of the problems caused by base encroachment since the emitter region does not overlap the extrinsic base region, as shown in Fig.7.5. In addition, a low-temperature epitaxial emitter is compatible with actual bipolar technology and potentially very advanced self-aligned schemes can be worked out. Since this material is doped in situ during deposition, its quality should be better than compensated crystalline Si, conventionally used in homojunction Si bipolar transistors. Therefore, we expect carrier lifetimes to be longer and bandgap narrowing to be less significant. The direct advantage would be an enhanced emitter injection efficiency. A typical output characteristics for a bipolar transistor having a Si emitter deposited at 250°C and recrystallized at 700°C for 20 min is depicted in Fig.7.6. The corresponding base Gummel number is $1.35 \cdot 10^{13}$ s/cm^4.

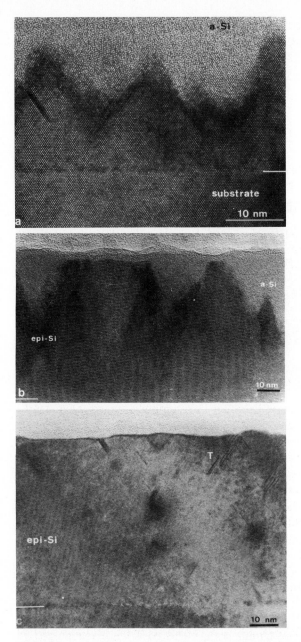

Fig.7.4a-c. (a) Cross sectional TEM photomicrograph showing that glow discharge growth of Si at 250°C on top of a clean c-Si interface results in a film which is partially epitaxial (under the deposition conditions followed in our experiments). The epitaxial region is defect free. (b) Cross sectional TEM photomicrograph (of the same film of Fig.7.4a) indicating that, in some regions, the epitaxial growth at 250°C proceeds up to surface. (c) Cross sectional TEM photomicrograph showing the same film of Figs.7.4a and b after annealing at 700°C for 20 minutes. The film becomes fully epitaxial with some twin defects appearing in the regions which were deposited amorphous

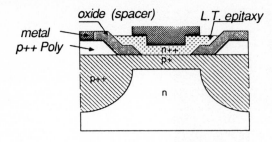

Fig.7.5. Schematic representation of an advanced self-aligned low temperature epitaxial-emitter bipolar transistor which uses an SiO₂ sidewall spacer between the emitter contact and the base contact

Low temperature epitaxial emitter

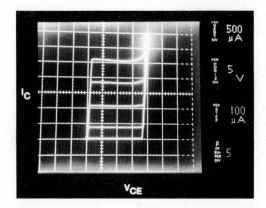

Fig.7.6. Curve tracer output characteristics (I_c-V_{ce}) of a recrystallized a-Si emitter bipolar transistor (the emitter layer is deposited at 250°C by glow discharge followed by a 700°C anneal for 20 min)

As expected, a current gain slightly larger than that of a conventionally implanted or diffused emitter transistor is observed.

Finally, it is expected that, with the progress achieved today in selective epitaxial growth, very advanced self-aligned epitaxial emitter (and base) structures could be realized.

7.4 Heterojunction Bipolar Transistors

The potential advantages of an emitter/base structure designed with the emitter having a wider bandgap than the base were recognized in 1957 by *Kroemer* [7.22]. One can make the height of the energy barrier for injection over a p-n junction different for holes and for electrons by choosing an n-type region and a p-type region with different bandgaps. This is the principle of a heterojunction first presented by *Shockley* in 1948 [7.23]. As shown in Fig.7.7, the barrier height for

118

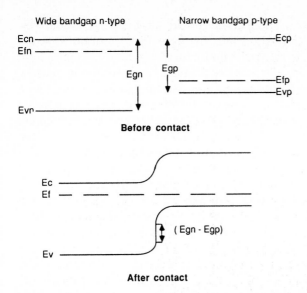

Fig.7.7. Band diagram at the heterojunction existing between n-type wide bandgap and p-type narrow bandgap materials. The materials are assumed to have the same electron affinity. All the bandgap difference appears as an offset in the valence band

hole injection is increased by the difference in the value of the bandgaps ΔE_G. Until the mid-seventies the idea could not be translated into practice due to the lack of suitable technology. In recent years, the advances in heterostructure technologies like Molecular Beam Epitaxy (MBE) and Metal Organic Chemical Vapor Deposition (MOCVD) have permitted an impressive development in heterostructure devices, especially in compound semiconductors like GaAs, GaAlAs, etc. Perfect or semi-perfect lattice match is one of the most important factors limiting the material system choice for heterojunctions. If one can overcome this problem by selecting appropriate materials, and succeed in realizing an npn transistor with emitter-base junction having a band diagram similar to that of Fig.7.7, hole injection from the base to the emitter is practically suppressed resulting in a tremendous increase in the emitter injection efficiency. In this case, the emitter efficiency is mainly determined by the recombination at the emitter-base heterointerface especially at low current levels. It should be mentioned that Fig.7.7 presents an ideal condition which is not met in practice. In fact, the energy bandgap difference (ΔE_G) is divided into a valence band offset (ΔE_v) and a conduction band offset (ΔE_c). The conduction band offset is mainly equal to the difference between the electron affinities of the n-type and p-type materials. An efficient junction for our application should have a large value for ΔE_v and a small value for ΔE_c, and therefore the condition imposed on the electron affinity is another important factor limiting the system choice. As shown in Fig.7.7, a wide bandgap emitter and a narrow bandgap base are conceptually equivalent and the choice of the name only depends on the system of materials forming the junction. For instance, if the substrate material is crystalline silicon, a wide bandgap emitter would use a material that has a bandgap wider than that of the silicon base. On the other hand, a narrow bandgap base would use a material that has a bandgap narrower than that of the silicon emitter.

Since in an efficient heterojunction (either wide-gap emitter or narrow-gap base), hole injection into the emitter is suppressed, the emitter delay caused by carrier storage in the emitter becomes negligible. Moreover, the emitter doping level can be orders of magnitude lower than in homojunction emitters without significantly increasing the minority carrier storage. The main advantages of low doping of the emitter are: 1) the possibility of heavily doping the base without inducing excessive emitter-base tunneling; this allows considerably thinner base widths without running into punchthrough conditions, and 2) reduced emitter-base junction capacitance. With a heavily doped base region, the extrinsic base doping profile can be made identical to that of the intrinsic base which eliminates extra steps required for the extrinsic base. In addition, two-dimensional problems related to base encroachment are eliminated. As a result of these advantages, it is expected that the ultimate speed performance of a silicon heterojunction bipolar transistor is superior to that of its homojunction counterpart.

The introduction of silicon HBTs in present or future Si IC fabrication has to face two main challenges. The first is to find an emitter that has a wide bandgap with a low bulk resistivity and which causes no contacting difficulties, or a base material that has a narrow bandgap. The second challenge resides in the compatibility of the material and of the process with state of the art silicon technology at minimum extra cost.

7.4.1 Wide-Bandgap Emitter Si Bipolar Transistors

The npn wide-bandgap-emitter Heterojunction Bipolar Transistors (HBT) based on GaAs technology are made with $Al_xGa_{1-x}As$ as the wide-bandgap-emitter material and GaAs as the material of the base and collector regions.

A number of different more or less successful approaches have been tried so far aiming at the realization of Si npn wide-bandgap-emitter HBTs. Among the wide-bandgap materials investigated we find:

1) *Semi-insulating polycrystalline silicon SIPOS*. Heavily phosphorus-doped SIPOS has a bandgap of 1.5eV. A large improvement in the current gain has been observed when using SIPOS emitters and it has been postulated that this improvement is obtained because the injected minority carriers see a barrier at the interface between the SIPOS emitter and the crystalline silicon base (emitter-base heterojunction) [7.24]. A high temperature anneal (900° - 1000°C for 1 hour) is necessary to transform the semi-insulating SIPOS layer into a semiconducting one with a resistivity of 100 Ωcm [7.24] and to obtain an emitter-specific contact resistance of the order of 10^{-6} Ωcm^2 [7.25]. During this high temperature step phosphorus atoms diffuse from the SIPOS film into the crystalline substrate (base). Therefore, the true emitter-base junction is situated in the crystalline silicon substrate and not at the silicon/SIPOS interface. It has been proved recently [7.25,26] that the large enhancement in the current gain is due to the presence of a thin interfacial oxide layer at the SIPOS/silicon interface. The interfacial oxide breaks up during the high temperature anneal, resulting in a greatly reduced emitter injection efficiency [7.26]. In conclusion, SIPOS emitters behave in a very similar manner to polysilicon emitters discussed in Sect.7.2.

2) *Hydrogenated amorphous silicon (a-Si:H)*. The system a-Si:H has a bandgap of 1.7eV. This material is deposited by glow discharge (plasma CVD) decomposition of silane (SiH_4) at a low temperature (250°C). In situ doping during deposition can be achieved by introducing a gas containing the dopant (e.g. PH_3) into the reaction chamber and results in a rather low- resistivity semi-conducting material (100Ωcm). In this case the emitter-base junction is truly situated at the silicon/a-Si interface since no high temperature anneal is required. A typical output characteristic (I_c-V_{ce}) of an a-Si:H emitter transistor fabricated in our laboratory is shown in Fig.7.8. The experimental values of the current gain of silicon homojunction transistors compared to those of a-Si:H emitter heterojunction transistors are plotted in Fig.7.9 as a function of the base Gummel number. One order of magnitude improvement in the current gain over that of conventional homojunction transistors has been observed [7.27,28]. As shown in Fig.7.10, the current gain strongly decreases at low current densities. This behaviour is most probably caused by significant recombination at the silicon/a-Si interface. Passivation of this interface is a very important step in the fabrication technique. The major drawback associated with a-Si emitters is their very high resistance. Although the a-Si:H film is heavily doped (10^{20}cm^{-3}), most of the P atoms are not electrically active resulting in a free carrier concentration

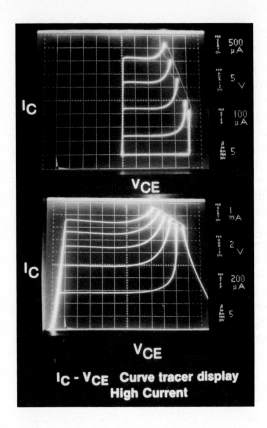

Fig.7.8. Curve tracer output characteristics (I_c-V_{ce}) of an n^+a-Si:H emitter bipolar transistor fabricated in our laboratory. The base Gummel number for this device is $1.35 \cdot 10^{13}$ s/cm^4

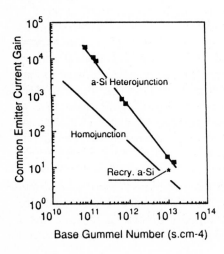

Fig.7.9. Experimental values of the maximum common emitter current gain β_{max} of the n^+a-Si:H emitter transistors versus the base Gummel number. The values for conventional homojunction transistors are also plotted for comparison. The star represents the value of β_{max} of the recrystallized a-Si emitter transistor

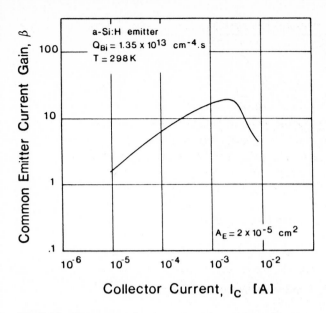

Fig.7.10. Common emitter current gain versus collector current for our first n⁺ a-Si:H emitter transistor

of the order of 10^{16} cm^{-3}. Also, the electron mobility in a-Si is at least two orders of magnitude smaller than in crystalline silicon. Experimental values of the a-Si emitter resistance are obtained from the I_b versus V_{ce} measurements. Typical results are shown in Fig.7.11. These curves are obtained by applying a constant current source to the base and measuring the collector-emitter voltage using a high-impedance voltmeter [7.29]. The measured specific resistance ranges between $4 \cdot 10^{-2}$ $\Omega \cdot$cm^2 (at 100°C) and $2\Omega \cdot$cm^2 (at -20°C). Since the bulk resistance contribution amounts to 10^{-3} $\Omega \cdot$cm^2, we deduce that most of the specific resistance is caused by the contact. The minimum specific resistance that one could obtain for n⁺ a-Si:H emitters is of the order of 10^{-4} $\Omega \cdot$cm^2. This value assumes a very efficient film doping (10$\Omega \cdot$cm), negligible contact resistance and a very thin layer (100nm). This is still too high for a VLSI bipolar transistor but can be acceptable for application in power integrated circuits.

3) *Hydrogenated amorphous silicon carbide (a-SiC:H)*. The system a-SiC:H has a bandgap of 1.8eV. It is deposited at low temperature (450°C) by glow discharge decomposition techniques. The gas mixture consists of silane (SiH$_4$), methane (CH$_4$), hydrogen (H$_2$), and a gas containing the dopant. The npn bipolar transistors using a-SiC:H as the emitter material have been fabricated and show an enhancement in the current gain over conventional homojunction transistors [7.30].

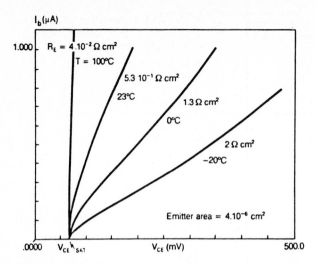

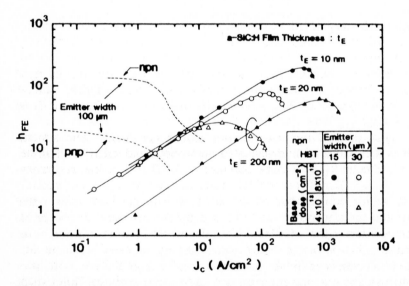

Fig.7.11. Determination of the emitter resistance from base current I_b versus collector to emitter voltage V_{ce} characteristics (*Getreu*'s method [7.29])

Fig.7.12. The common emitter current gain of a-SiC:H emitter HBT versus collector current density [7.30]

The emitter Gummel number of this transistor is estimated to be 2.10^{14} s/cm^4. The dependence of the maximum current gain on the width of the emitter stripe shown in Figs.7.12 and 13 indicates the presence of significant emitter crowding. In simple terms, a-SiC:H emitters suffer from very high emitter resistance. In addition, the

124

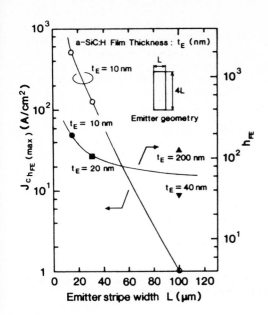

Fig.7.13. Dependence of the collector current density on an a-SiC:H HBT at maximum current gain on the emitter stripe width [7.30]

current gain is very small at low current densities which indicates the presence of large recombination currents at the a-SiC:H/Si interface. The pnp transistors with p^+a-SiC:H emitters have also been tested [7.31] and show a current gain 30 times smaller than that of the npn transistors. The difference seems to confirm the band diagram picture of Fig.7.7 for the a-SiC:H/Si system with ΔE_v = 570 meV and ΔE_c = 130 meV [7.30].

4) *Hydrogenated micro-crystalline silicon (μc-Si:H)*. The system μc-Si:H is a mixture of hydrogenated amorphous and micro-crystalline phases of silicon. It has a large bandgap (1.5-1.9 eV), a low resistivity (0.009-0.22 $\Omega \cdot cm$) and a very low minority carrier diffusion coefficient. It is therefore expected to be a serious candidate for wide bandgap emitter transistors. μc-Si:H is deposited by glow discharge at low temperature (250°-450°C). The deposition parameters such as temperature, RF power, pressure, etc. must be adjusted to avoid the deposition of purely a-Si. As shown in Fig.7.14, npn HBTs with μc-Si:H emitters show a current gain one order of magnitude larger than that of homojunction transistors [7.32]. The non-ideal slope of the base current versus V_{be} displayed in Fig.7.15 and the dependence of the current gain on the collector current density plotted in Fig.7.16 clearly indicate the presence of interface recombination currents. The current gain degradation at low current densities is, however, reduced in comparison to the results of a-Si:H or a-SiC:H emitters. It is also possible to transform a-Si:H material to μc-Si material by thermal annealing at temperatures near to 600°-700°C. However, μc-Si obtained by this

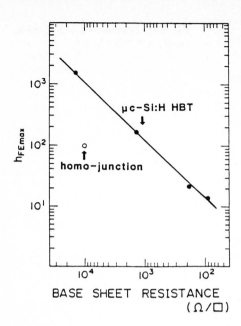

Fig.7.14. Maximum common emitter current gain of μc-Si:H emitter HBT as a function of intrinsic base sheet resistance. Typical current gain of homojunction Si transistor is also indicated [7.32]

method is not hydrogenated and has the tendency align itself epitaxially to the substrate. If we consider a transistor having an emitter material obtained by thermal annealing of an a-Si:H film at 700°C, it is not possible to confirm whether it is a μc-Si emitter HBT or a low temperature epitaxial bipolar transistor. In fact, the TEM photograph of Fig.7.4c indicates that the resulting material can be purely epitaxial. Poor reproducibility is therefore the major difficulty encountered so far with the use of μc-Si.

5) *Crystalline silicon carbide (β-SiC).* The system β-SiC is expected to play an important role in electronic devices in general, and especially those operated at high temperatures. For our application this material seems very promising since it has a wide bandgap of 2.2 eV, a controllable electrical conductivity and is thermally stable. A large lattice mismatch of about 20% exists between SiC and Si. Due to this large mismatch a large number of misfit dislocations is created at the heterojunction and in the SiC layer. These dislocations are, however, not electrically active and therefore do not act as effective recombination centers. The epitaxial growth of β-SiC can be carried out by chemical vapor deposition at atmospheric or low pressure (APCVD or LPCVD) using C_3H_8-SiH_4-H_2 or C_3H_8-$SiHCl_3$-H_2 reaction gas systems. Very good quality CVD epitaxial growth on (100) crystalline silicon has been reported [7.33,34]. However, due to the fact that the growth of β-SiC by CVD is carried out at high temperatures (1000°-1300°C), emitter-base dopant interdiffusion might occur

126

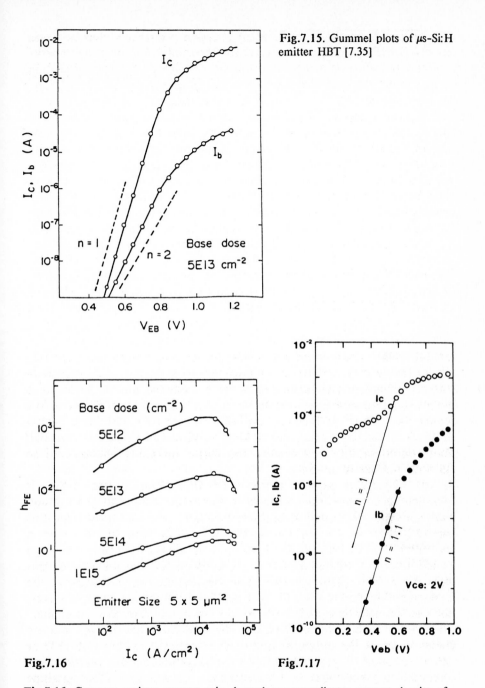

Fig.7.15. Gummel plots of μs-Si:H emitter HBT [7.35]

Fig.7.16

Fig.7.17

Fig.7.16. Common emitter current gain dependence on collector current density of μc-Si:H emitter HBT [7.32]

Fig.7.17. Gummel plots of β-SiC emitter HBT [7.35]

127

causing a shift in the true electrical junction position. Molecular beam epitaxy (MBE) [7.19b] can be used for growing β-SiC at low temperature and therefore is a more appropriate technology for the fabrication of a true emitter-base heterojunction. Very recent reports [7.35] demonstrated the successful operation of a single-crystal SiC/Si heterojunction bipolar transistor. Its common emitter current gain is 800 for a base Gummel number of 10^{11} s/cm^4. As shown in Fig.7.17, the base current has a non-ideality factor of 1.1 which suggests that it is mainly dominated by diffusion, which confirms the idea that the interface dislocations are not effective recombination centers. The observed non-ideal behaviour of the collector current is not intrinsic to the device but is caused by collector to emitter punchthrough due to the low base doping of the transistor under test. Although the observed current gain is comparable to that of an identical base homojunction transistor, the SiC emitter has a free carrier density of only 10^{19} cm^{-3}. This low free-carrier concentration also causes an emitter resistance problem. Indeed, the SiC layer resistivity is estimated to 6.3·10^{-2} Ω·cm which is at least 10 times larger than that of polysilicon emitter layers. It is expected that epitaxial SiC layers with larger free carrier concentration, hence lower resistivity and smaller base current can be achieved by optimizing the epitaxial growth condition.

6) *Gallium phosphide (GaP)*. GaP has a bandgap of 2.26 eV and is lattice matched to silicon to within 0.4%. Epitaxial growth of GaP on silicon can be carried out by means of CVD [7.36] or MBE [7.37,38]. In the GaP on Si system, interdiffusion of emitter and base dopant atoms (Si from the base to the epitaxial emitter and P from the emitter to the base) is highly probable especially in high temperature CVD deposition. Interface mixing of dopants results in an interface neutrality situation which is detrimental for the emitter injection efficiency. This problem is less significant with low temperature deposition techniques such as MBE, which is therefore highly recommended. The best value reported so far for the emitter efficiency of a Si HBT with a GaP emitter deposited on Si by MBE is 0.9 [7.38], corresponding to a common emitter current gain of 9. It is speculated that a deep conduction band notch occurs in the band diagram at the GaP/Si interface [7.38]. Therefore, the current gain could be limited by tunneling currents at the interface. However, with a maximum base doping of 10^{19} cm^{-3} and a base width of 0.18 μm, the base Gummel number is estimated to be in the range of 10^{13}-10^{14} s/cm^4. For such a transistor, a current gain of 9 is quite promising and represents an improvement compared to homojunction Si transistors.

7.4.2 Narrow-Bandgap Base Si Bipolar Transistors

A silicon heterojunction emitter-base diode can be obtained by using an emitter with a wider bandgap than silicon, as explained in the previous subsection, or by using a silicon emitter on top of a narrow-bandgap base. Such a narrow-bandgap base transistor is, in fact, a double heterojunction transistor, since it has a silicon collector. The base-collector and base-emitter junctions have identical band-offsets. Several advantages of double heterojunction bipolar transistors have been described by *Kroemer* [7.22]. The most important ones are: 1) the interchangeability of emitter and collector due to the elimination of the collector-emitter offset voltage, and 2) the suppression of holes in the collector eliminating charge storage under saturation conditions. Other attractive features of narrow-bandgap base transistors are the reduced turn-on voltage and smaller power dissipation. In addition, the narrow-bandgap base structure does not suffer from emitter contact resistance problems typical for most wide bandgap materials compatible with silicon technology.

The most obvious way to obtain a narrow-bandgap base silicon transistor is to alloy the silicon base region with germanium resulting in the $n^+Si/p^+Ge_xSi_{1-x}/n$ Si structure described in Fig.7.18. This alloy can be obtained by MBE, ultrahigh vacuum chemical vapor deposition or simply by implanting silicon with a heavy dose of Ge ions. Although the Ge_xSi_{1-x} base layer is not lattice matched to the underlying collector, it is well known [7.39,40] that Ge_xSi_{1-x} layers can be grown pseudomorphically on silicon if the layers are thinner than the critical thickness. If such a p-type Ge_xSi_{1-x} base layer is overgrown by an n-type doped emitter layer, a narrow-base transistor is formed. For x = 0.2, a bandgap difference of about 170 meV between emitter and base can be expected with $\Delta E_c = 20$ meV and $\Delta E_v = 150$ meV, respectively [7.41]. These values are based on the fact that the Ge_xSi_{1-x} base is pseudomorphically strained and has the lattice constant of the Si substrate. The collector and emitter layers, on the other hand, are unstressed and cubic. The proposed energy diagram of the $Si/Ge_{0.2}Si_{0.8}$ emitter-base junction is shown in Fig.7.19. Since ΔE_v is large and ΔE_c is small, an efficient heterojunction emitter with almost

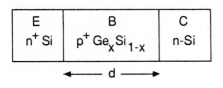

Fig.7.18. Bipolar transistor formed by a double heterojunction structure. The emitter and the collector are made of crystalline Si and the base is made of a narrower bandgap materials (Si-Ge alloy). For lattice match, the base width d must allow a commensurate growth

129

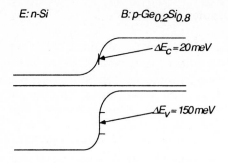

E: n-Si   B: p-Ge$_{0.2}$Si$_{0.8}$

$\Delta E_C = 20\,meV$

$\Delta E_V = 150\,meV$

Fig.7.19. Band diagram at the n-Si emitter/p-Ge$_{0.2}$Si$_{0.8}$ base junction (also at the collector-base junction). Note that most of the difference in the energy bandgaps (170 meV) appears as a valence band offset (150 meV)

no spike on the conduction band edge is formed. With x = 0.2 the critical base thickness for obtaining pseudomorphic growth of the base layer and an unstressed emitter is 200 nm. This value is sufficiently large to avoid emitter-to-collector punchthrough and to yield a small base-sheet resistance if sufficiently large base doping levels are used. The strained base layer offers another advantage to the device operation. Indeed, due to the modification of the conduction-band structure, the electron effective mass becomes smaller, which enhances the electron mobility in the base.

The practical operation of pseudomorphic base heterojunction bipolar transistors has recently been demonstrated in III-V technology with a AlGaAs/GaInAs emitter-base heterojunction [7.42]. It is predicted that the performance of Si/SiGe HBTs will be comparable to that of GaAs HBTs with a cut-off frequency and a maximum frequency reaching 20 and 40 GHz, respectively [7.43]

As pointed out, true Si heterojunction bipolar transistors can be processed only at low temperature to avoid emitter-base dopant interdiffusion. Therefore, molecular beam epitaxy and ultrahigh vacuum chemical vapor deposition epitaxy are the most appropriate processing techniques for this application.

Ge$_x$Si$_{1-x}$ base HBTs have been grown recently by MBE [7.44, 45]. Due to the presence of the interface recombination currents, the improvement achieved in the current gain with Ge$_{0.3}$Si$_{0.7}$ base is limited to 2.5-15 times its value with a Si base [7.44]. The small bandgap in the base results in an enhanced collector current. The ratio of the alloyed-base collector current to the Si-base collector current is determined experimentally and plotted as a function of the percentage of Ge in Fig.7.20 [7.45]. The same ratio is plotted for a 12% Ge content (Ge$_{0.12}$Si$_{0.88}$) versus inverse of temperature in Fig.7.21 and the slope exhibits an energy difference of 59 meV which represents the amount of bandgap reduction in the base [7.45]. The advantage of heterojunction transistors for low temperature operation has also been demonstrated [7.45]. While a 5.7 times increase in the collector current is obtained at room temperature with a Ge$_{0.12}$Si$_{0.88}$ base, a 1000 times

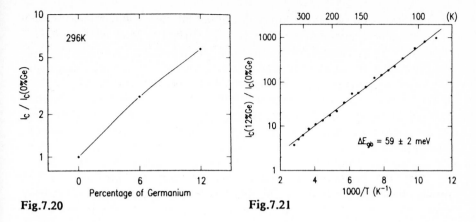

Fig.7.20

Fig.7.21

Fig.7.20. Dependence of collector current on Ge concentration at 296 K for a 0.1 μm base transistor. Collector current values have been normalized to the value for a Si-base transistor at the same bias V_{be} = 0.5 V [7.45]

Fig.7.21. The ratio of the collector current of a $Ge_{0.12}Si_{0.88}$ base heterojunction transistor to that of a Si base transistor at V_{be} = 0.5 V versus inverse of temperature [7.45]

increase is achieved at 90 K which results in a higher current gain at 90 K than at room temperature [7.45].

7.5 Conclusions

In scaled silicon bipolar transistors the emitter plays a crucial role. Although polysilicon emitters allow super-self-aligned structures, their use in scaled transistors with ultra-shallow junctions has several drawbacks. A major problem is the high emitter series resistance. A high-temperature treatment is necessary to reduce the emitter resistance, resulting in diffusion of the dopant from the polysilicon into the underlying monocrystalline silicon. This outdiffusion causes two main problems: (i) a lower emitter efficiency, and (ii) a large minority carrier charge storage reducing the cut-off frequency of the device.

Epitaxial emitters are potentially superior to polysilicon emitters if the epi-growth is performed at sufficiently low temperatures to prevent emitter dopant outdiffusion. Novel experimental results indicate that heavily doped n+ epitaxial layers with reasonable quality can be grown without exceeding 700°C using high throughput glow-discharge equipment.

A further step is the use of heterojunction emitter-base diodes. Such structures eliminate the emitter stored charge without requiring large emitter doping levels, relax the punchthrough/tunneling trade-off and can potentially reduce the two-dimensional base encroachment problems. With these devices the base can be doped heavily without compromising the transistor action and therefore very small base resistance can be achieved. Most wide bandgap emitters on a silicon base suffer, however, from inherent emitter contact resistance problems strongly reducing their potential use of VLSI applications. For this reason narrow bandgap base transistors seem the most promising silicon heterojunction approach. The $Si/Ge_x Si_{1-x}$ emitter-base junction is a very attractive candidate. Preliminary results indicate that this alloyed base has the potential to significantly improve the performance of bipolar transistors and make feasible bipolar operation at low temperatures.

References

7.1   J.M.C. Stork, R.D. Isaac: IEEE Trans. ED-30, 1527 (1983)
7.2   P.M. Solomon: Proc. IEEE 70, 489 (1982)
7.3   C.T. Chuang, D.D. Tang, G.P. Li, E. Hackbarth: IEEE Trans. ED-34, 1519 (1987)
7.4   M. Takagi, K. Nakayama, C. Tevada, H. Kamioko: J. Jpn. Soc. Appl. Phys. 42, 101 (1972)
7.5   J. Graul, A. Glasl, H. Murrmann: IEEE J. SC-11, 491 (1976)
7.6   H.C. De Graaff, J.G. De Groot: IEEE Trans. ED-26, 1771 (1979)
7.7   T.H. Ning, R.D. Isaac: IEEE Trans. ED-27, 2051 (1980)
7.8   A.A. Eltoukhy, D.J. Roulston: IEEE Trans. ED-29, 1862 (1982)
7.9   P. Ashburn, B. Soerowirdjo: IEEE Trans. ED-31, 853 (1984)
7.10   T.H. Ning, R.D. Isaac, P.M. Solomon, D.D. Tang, H.N. Yu, G.C. Feth, S.K. Wiedmann: IEEE Trans. ED-28, 1010 (1981)
7.11   E. Crabbé, S. Swirhun, J. del Alamo, R.F.W. Pease, R.M. Swanson: IEDM Tech. Dig. (1986) p.28
7.12   G.L. Patton, J.C. Bravman, J.D. Plummer: IEEE Trans. ED-33, 1754 (1986)
7.13   M.Y. Ghannam, R.W. Dutton: Appl. Phys. Lett. 51, 611 (1987)
7.14   W.R. Burger, R. Reif: IEEE EDL-6, 652 (1985)
7.15   T.N. Nguyen, D.L. Harame, J.M.C. Stork, F.K. LeGoues, B.S. Meyerson: IEDM Tech. Digest 304 (1986)
7.16   B.S. Meyerson: Appl. Phys. Lett. 48, 797 (1986)
7.17   K. Baert, J. Nijs, J. Symons, J. Vanhellemont, W. Vandervorst, R. Mertens: Appl. Phys. Lett. to be published
7.18   P. De Pauw: Ph.D. Dissertation, K.U. Leuven, Belgium (1984)
7.19   Y. Ota: J. Electrochem. Soc. 126, 1761 (1979)
       M.A. Herman, H. Sitter: *Molecular Beam Epitaxy*, Springer Ser. Mat. Sci., Vol.6 (Springer, Berlin, Heidelberg 1988)
7.20   M.G. Duchemin, M.M. Bonnetand, M.F. Koelsch: J. Electrochem. Soc. 125, 637 (1980)
7.21   R.G. Friezer: J. Electrochem. Soc. 115, 401 (1968)
7.22   H. Kroemer: Proc. IEEE 70, 13 (1982)
7.23   W. Shockley: U.S. Patent 2 569 347, 1951
7.24   N. Oh-uchi, H. Hayashi, H. Yamoto, T. Matsushita: IEDM Tech. Dig. (1979) p.522

7.25 E. Yablonovitch, T. Gmitter: IEEE EDL-6, 597 (1985)

7.26 R.M. Swanson, Y.H. Kwark: Proc. U.S.- Belgium Joint Seminar on "New Developments in the physics of homo- and heterojunctions", IMEC Leuven, Belgium (1986); and Solid St. Electronics 30, 1121 (1987)

7.27 M. Ghannam, J. Nijs, R. Mertens, R. de Keersmaecker: IEDM Tech. Dig. (1984) p.746

7.28 J. Symons, M. Ghannam, J. Nijs, A. Van Ammel, P.De Schepper, A. Neugroschel, R. Mertens: Proc. U.S.- Belgium Joint Seminar on "New developments in the physics of homo- and heterojunctions", IMEC, Leuven, Belgium (1986); and Solid St. Electronics 30, 1143 (1987)

7.29 I. Getreu: *Modeling the Bipolar Transistor* (Tektronix Inc., Beaverton, OR 1976)

7.30 K. Sasaki, S. Furukawa, M.M. Rahman: IEDM Tech. Dig. (1985) p.294

7.31 K. Sasaki, M.M. Rahman, S. Furukawa: IEEE EDL-6, 311 (1985)

7.32 H. Fujioka, S. Ri, K. Takasaki, K. Fujino, Y. Ban: IEDM Tech. Dig. (1987) p.190

7.33 S. Nishino, J.A. Powell, H.A. Will: Appl. Phys. Lett. 42, 460 (1983)

7.34 Y. Furumura, M. Doki, F. Mieno, M. Maeda: Trans. IECEJ J69-C, 705 (1986)

7.35 T. Sugii, T. Ito, Y. Furumura, M. Doki, F. Mieno, M. Maeda: IEEE EDL-9, 87 (1988)

7.36 T. Katoda, M. Kishi: J. Electron. Mater. 9, 783 (1980)

7.37 H. Kawanami, T. Sakamoto, T. Takahashi, E. Suzuki, K. Nagai: Jpn. J. Appl. Phys. 21, L68 (1982)

7.38 S.L. Wright, H. Kroemer, M. Inada: J. Appl. Phys. 55, 2916 (1984)

7.39 H. Jorke, H.J. Herzog: J. Electrochem. Soc. 133, 998 (1986)

7.40 R. People, J.C. Bean: Appl. Phys. Lett. 47, 322 (1985)

7.41 R. People, J.C. Bean: Appl. Phys. Lett. 48, 538 (1986)

7.42 P.M. Enquist, L.R. Ramberg, F.E.Najjar, W.J. Schaff, L.F. Eastman: Appl. Phys. Lett. 49, 179 (1986)

7.43 C. Smith, A.D. Welbourn: Proc. IEEE 1987 Bipolar Circuits and Technology Meeting (1987) p.57

7.44 T. Tatsumi, H. Hirayama, N. Aizaki: Appl. Phys. Lett, 52, 895 (1988)

7.45 G.L. Patton, S.S. Iyer, S.L. Delage, S. Tiwari, J.M.C. Stork: IEEE EDL-9, 165 (1988)

8. Molecular Beam Epitaxy of Silicon-Based Bipolar Structures

E. Kasper, P. Narozny, and K. Strohm
AEG-Research Center
D-7900 Ulm, Fed. Rep. Germany

It is possible to grow a thin single crystalline layer on top of a substrate which delivers mechanical support and orientation information to the film. This process is called epitaxy. Traditionally, bipolar circuits have been formed in epitaxial layers. The standard method of production of epitaxial layers is Chemical Vapour Deposition (CVD). Recently, Molecular Beam Epitaxy of silicon and silicon-based heterostructures (Si-MBE) has gained attention for research into advanced structures and devices. About fifty laboratories around the world are now working in this field. A rigorous treatment of the method, the equipment, the applications and a complete bibliography are given in [8.1].

The first investigations were performed with home-made equipment. Now, however, commercial manufacturers (e.g., Atomika, Anelva, Perkin-Elmer, Riber, Vacuum Generators, Varian) are also offering equipment. Nonetheless, the equipment market is oriented towards the research scene. First moves towards production equipment stem from cooperative projects between equipment manufacturers and the electronics industry. Design principles of maximum simplicity, batch processing on rotating plates, process control by in situ monitoring, and automatic wafer transfer in ultrahigh vacuum from casette-type magazines into the growth chamber, mark the way to a production equipment [8.1].

8.1 Strengths of Si-MBE

In MBE, molecular beams of the matrix and dopant elements (Fig.8.1) are directed onto the heated surface of the substrate. The adsorbed atoms may be incorporated in the growing crystal, or they may desorb or segregate. The properties of the adatom population are mainly a function of surface orientation and temperature if contamination of the surface is avoided by properly chosen ultrahigh vacuum (UHV) conditions. Si-MBE has greatly extended our knowledge about the atomistic nature of the growth process, and possibilities for the growth of sili-

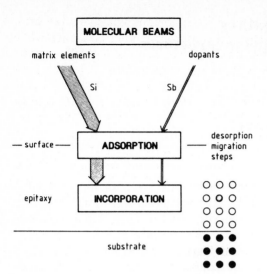

Fig.8.1. Basic scheme of deposition from molecular beams. Adsorbed atoms can incorporate, desorb or segregate

con–based materials have expanded in a dramatic manner. Some examples relevant for bipolar processing are given in the following sections.

8.1.1 Low Processing Temperatures

Less than a decade ago, the majority of the semiconductor community believed that single crystalline growth of silicon was restricted to a rather small temperature regime below the melting point. From roughly 750 K to 1200 K a polycrystalline phase, and below that temperature regime an amorphous phase was assumed. MBE proved that the single crystalline phase on (100) substrates extends over more than 1000 K below the melting point. At temperatures below 400 K to 550 K [8.2] the growth mode switches to an amorphous structure. The polycrystalline phase is only produced when the necessary orientation information is missing, either due to the intentional choice of an amorphous substrate (glass, oxide), or by unintentional contamination of the surface. The latter was the source of the old results, now identified as untypical for clean silicon growth [8.3]. Such clean surfaces allow high surface diffusivity of adatoms with diffusion barriers below 1 eV [8.3]. Surface step motion and adatom diffusion is strongly hampered by even a small contamination far below one monolayer [8.4]. From these MBE experiments a clear message follows: the tendency of nature is to realize the single crystalline phase, which is obviously the state of lowest energy. For growth on a Si substrate, only very low growth temperatures or surface contamination caused by improper experimental conditions can interfere with this tendency.

The most important sources of surface contamination are the preparation of the substrate surface itself, and the sticking of contaminating gases during the growth process. The most popular Si-MBE growth sequence contains a substrate preparation with a thin (0.5 to 1 nm) protecting oxide, an in situ thermal cleaning step and the growth cycle in UHV surroundings. Growth proceeds under standard MBE conditions between 800 K and 1050 K (Fig.8.2). Only a slight modification of the temperature programme is needed for a Solid Phase Epitaxy (SPE) version of the growth process. Deposition of an amorphous layer takes place at room temperature followed by a recrystallization of the layer at about 850 K. Dopant levels above the solid solubility limit can be obtained with this process modification which is called solid phase MBE [8.5].

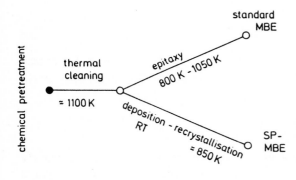

Fig.8.2. Three-stage process sequence of Si-MBE. Chemical precleaning, in situ heating, epitaxy

8.1.2 Abrupt Dopant Transitions

With chemical vapour deposition (CVD), transfer of dopant material from the substrate or highly doped regions (buried layer) to the epitaxial film is a serious problem (autodoping). Autodoping of the epitaxial film by transfer of dopant material from the substrate is completely avoided in Si-MBE because of the low process temperatures. Doping profiles [8.6] of an n-epitaxial layer on a p-wafer with n^+-buried layer zones are shown in Fig.8.3. The n-doping level of the epitaxial layer is not degraded by the existence of the buried layer regions, and the interfaces are completely abrupt both to the lightly doped p^--substrate and to the highly doped n^+-buried layer (the small interface transition in Fig.8.3 is caused by the limited resolution of the measurement method).

In the standard bipolar IC process, base and emitter regions are created by diffusion and/or implantation into the n^--epitaxial layer. The effective width d_{eff} of the remaining n^--region (collector) is given

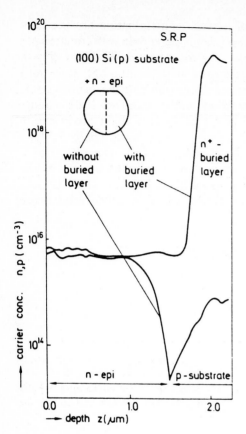

Fig.8.3. Doping profile of a n-type Si-MBE layer on a p⁻ substrate with n⁺-buried layer regions [8.6]. Spreading resistance probe (SRP) measurements demonstrate abrupt junctions without autodoping

by the epitaxial thickness d minus the base and emitter width and minus spread out of the buried subcollector caused by autodoping and post-epitaxial diffusion. With this standard bipolar process, the full strength of MBE cannot be exploited because base/emitter definition and post-epitaxial diffusion will result in a smear out of the abrupt MBE profiles. We have used this standard bipolar process [8.6] for fabrication of GHz frequency dividers where, except for the Si-MBE layer, only a commercial process line was used. The successful fabrication of bipolar IC's with high yield demonstrated the crystallographic quality of the MBE material. The exact definition of the epilayer thickness allowed an optimization of the operating frequency for a given design and process sequence (Fig.8.4). In the given example the transit frequency f_T of the bipolar transistor increased with decreasing epitaxial thickness d down to $d_s = 1$ μm. With further decreasing d the transit frequency f_T decreased because of the vanishing collector region ($d_{eff} \rightarrow 0$).

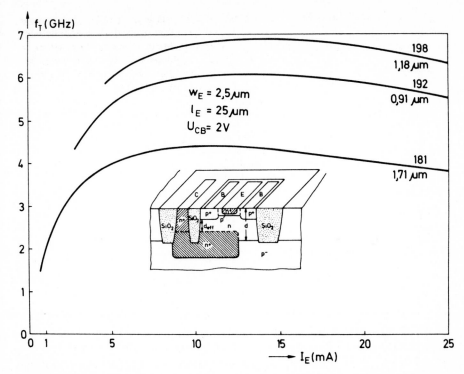

Fig.8.4. Transit frequency f_T versus collector current I_c of integrated bipolar transistors with different epitaxial layer thicknesses. Inset: Cross-section of the transistor structure

8.1.3 Complex Doping Profiles

Usually emitter and base profiles are defined by post-epitaxial implantation or diffusion steps. In principle, MBE offers a one-step definition of collector, base and emitter profiles. An example [8.7] of an npn$^+$ structure grown by Si-MBE in a one-step deposition is shown in Fig.8.5. Clearly seen is the complex vertical profile with sharp junctions and well-defined doping levels. Nonetheless, a real test of such complex doping structures in integrated circuits remains to be performed, mainly for two reasons. Firstly, modified techniques for lateral definition still have to be developed, and secondly, Si-MBE doping techniques for advanced bipolar transistors are still in their infancy.

In a simple picture of MBE, doping proceeds by simultaneous co-evaporation of matrix and dopant elements. Indeed, in III/V-MBE this simple picture was realistic for some dopant elements. But Si-MBE was challenged by a situation completely different from the simple picture, manifested by severe surface segregation and clustering of the dopants.

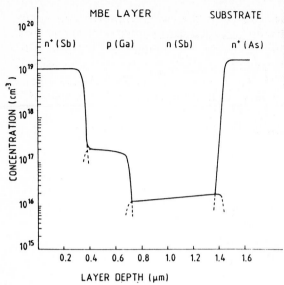

Fig.8.5. Complex doping profile realized in a one-step Si-MBE process. The n-p-n$^+$-doping profile with abrupt junctions and defined doping levels [8.7]

A lot of effort was invested in overcoming this challenging problem. Si-MBE now offers solutions to many doping problems, and rapid progress continues to be made. This phase was not only fruitful for Si-MBE itself, but also for the development of sophisticated doping methods [8.1] such as low energy ion implantation, secondary implantation, pre-build-up and flash-off of dopant adlayers, and δ-doping. For physical reasons Sb and Ga were the primary n- and p-dopant elements in Si-MBE. In recent years, B from newly designed effusion cells has been replacing Ga as the primary p-dopant, and low energy As$^+$ ions are competing with Sb as the primary n-dopant.

It is now possible with MBE to realize almost any doping profile. We hope to encourage device designers and manufacturers to demand the realization of doping profiles that are optimized for their application.

8.1.4 Lateral Isolation

Lateral isolation by p-n junctions or oxides needs additional process steps critical with respect to mask alignment and heat budget. An epitaxy concept, which needs growth of single crystalline collector and base/emitter islands within oxide windows, opens up an alternative route to advanced bipolar circuits. Although selective epitaxy, which means growth within the windows without growth on the oxide, is also possible in Si-MBE at certain temperatures, the main emphasis is directed towards differential epitaxy [8.8]. With differential epitaxy,

140

single crystalline islands are grown in the oxide windows surrounded by an oxide layer covered with fine-grained poly-Si with unique properties. This poly-Si is semi-insulating at doping levels below 10^{18} cm^{-3} but switches to good conductivity [8.9] (similar to the conductivity of single crystalline Si of the same doping level) at higher doping levels ($>10^{19}$ cm^{-3}). It was demonstrated (Fig.8.6) that differential MBE can be used for the fabrication of integrated GHz-transistors [8.8] and for the realization of resistor structures by ion implantation in the poly-Si.

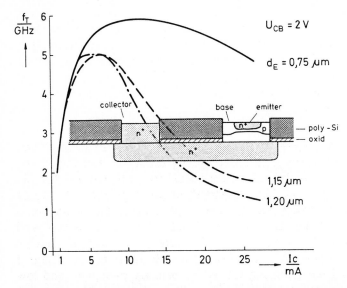

Fig.8.6. Lateral isolation of an integrated bipolar transistor by differential epitaxy [8.8]. Transit frequency f_T versus collector current of the transistor. Inset: Transistor structure in epitaxial islands separated by semi-insulating poly-Si on oxide

8.1.5 Heteroepitaxy

There is little doubt that future microelectrons will be based mainly on Si substrates. The reasons for this are physical (thermal conductivity, oxide interface), technical (quality, hardness, size), economical (resources) and ecological (silicon is one of the relatively few harmless elements).

Silicon-based heterostructures promise to improve the performance of existing device types and to enable tailoring of new material and device properties [8.10]. A key problem with heterostructures is the degradation of material quality and interface/surface morphology by lattice mismatch, thermal expansion differences and chemical forces.

141

The most promising heterostructure systems are listed in Table 8.1. They include silicon-semiconductor, silicon-metal and silicon-insulator systems.

Particularly for the semiconductor heterostructures (SiGe/Si and GaAs/Si), there is considerable lattice mismatch (up to 4%). For such mismatched systems we should clearly distinguish two thickness regimes. For small layer thicknesses (below a critical thickness t_c) lattice accommodation is obtained by film strain (termed pseudomorphic growth or commensurate or strained layer growth). For thicker layers accommodation is mainly obtained by misfit dislocations (termed incommensurate or relaxed layer growth). For a more general approach to heterostructures or superlattices, a strain adjustment by an incommensurate buffer layer is needed [8.10] but, for the near future, for integration with Si-IC's, commensurate hetero-layers will be preferred because of the superior crystal quality.

Table 8.1. Silicon-based heterostructure systems of technical importance

Si semiconductor	Si metal	Si insulator
SiGe, Ge	$NiSi_2$	CaF_2
GaAs	$CoSi_2$	

8.2 Silicon Monolithic Millimeter Wave Integrated Circuits (SiMMWICs)

The millimeter wave range is defined as the frequency region from 30 to 300 GHz. This corresponds to wavelengths from 10 mm to 1 mm, as indicated by the term "millimeter wave". Microelectronics is just beginning to enter this frequency range. Two approaches are possible: The common one starts from low frequencies and moves slowly towards higher frequencies by scaling and improving current devices and technology. The other approach directly explores a technology at higher frequencies. 90 GHz is a good choice. At 90 GHz there is an atmospheric window with low atmospheric absorption (0.1dB/km), which enables numerous technical applications. Furthermore the wavelengths at 90 GHz are small enough for the integration of complete receiver and transmitter chips including antennas. These antennas allow high energy focussing and enable sharp alignment and imaging. Therefore communication links and completely new sensors can be manufactured for industrial test methods, production automation, traffic control, identification and security surveillance systems.

8.2.1 Monolithic Integration of Active Devices with Passive Components on a Semi-Insulating Substrate

In millimeter wave circuits the usual transmission line is the microstrip line. A microstrip transmission line consists of a strip conductor and a ground plane separated by a dielectric medium (Fig.8.7a). Since field lines between the strip and ground plane are not entirely confined to the substrate (Fig.8.7b), the mode propagating along the strip is not purely transverse electromagnetic (TEM) but quasi-TEM [8.11].

The characteristics of microstrip lines may be described by three parameters: microstrip line wavelength, characteristic impedance and attenuation. The attenuation constant, α, is one of the most important characteristics of any transmission line. There are two sources of dissipative losses in a microstrip circuit: conductor loss and substrate dielectric loss. Dielectric loss depends on the substrate resistivity, but is nearly independent of frequency, whereas conductor losses are independent of substrate, but increase strongly with frequency (skin-effect, radiation). This is shown in Fig.8.8, where the substrate losses for different frequencies are shown as a function of line impedance. For high resistivity silicon substrates ($\rho > 2000\,\Omega\cdot\mathrm{cm}$) it can be concluded that, in the millimeter wave range (>30GHz), the conductive loss dominates and that the substrate loss does not contribute significantly to the total loss.

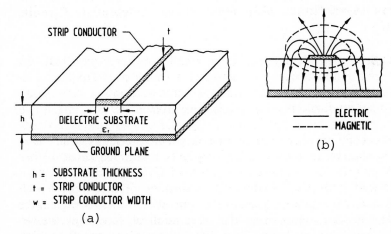

h = SUBSTRATE THICKNESS
t = STRIP CONDUCTOR
w = STRIP CONDUCTOR WIDTH

(a)

Fig.8.7a,b. A microstrip transmission line consists of a strip conductor and a ground plane separated by a dielectric medium (a). Since not all field lines are contained in the substrate, the propagation mode is quasi-TEM (b)

143

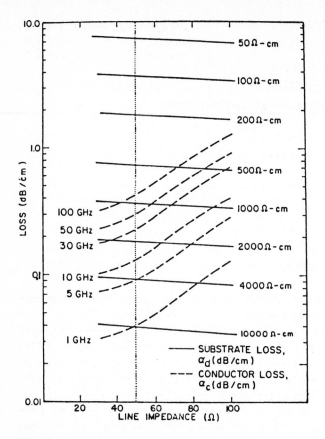

Fig.8.8. Substrate loss for different substrate resistivities and conductor loss for different frequencies as a function of line impedance

This is clearly demonstrated in Fig.8.9, where the total loss (substrate and conductor loss) is shown as a function of frequency for 50 Ω microstrip lines on Si, GaAs and ceramic substrates. With respect to microstrip attenuation, high resistivity silicon is not inferior to GaAs.

The concept of silicon monolithic microwave circuits is not new. Its origin goes back to 1964 to a US founded programme based on silicon technology [8.12]. Measurements of microstrip lines on 1400 Ω·cm silicon showed promising attenuation values, as predicted by theory [8.13]. However, it was found that insulating silicon lost its high resistivity characteristics during high-temperature processing steps. *Battershall* and *Emmons* [8.14] report, that 800 Ω·cm p-Si undergoes a resistivity change to 1-10 Ω·cm n-Si after approximately 6 hours at 1100°C. This inversion problem, resulting from high temperature pro-

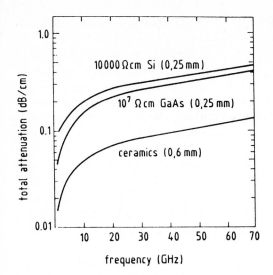

Fig.8.9. Total loss of 50 Ω microstrip lines on different substrates as a function of frequency

cessing sequences, led to a deterioration of the quality of the substrates in a manner unacceptable for microwave circuitry [8.15].

Therefore it was generally believed that the processing temperature for the fabrication of active devices should not exceed 800°C. This was first fulfilled in 1981 by a group at RCA using ion implantation and laser annealing [8.16]. At the same time AEG had developed a low-temperature deposition technique, the silicon molecular beam epitaxy (Si-MBE) for the growth of active, thin monocrystalline films with abrupt junctions and high vertical resolution at temperatures between 550°C and 750°C [8.6] and was testing X-ray lithography, a new lithographic method with high lateral (submicron) and high topographic resolution [8.17]. These two VLSI techniques seemed to be promising for the silicon monolithic millimeter wave technology. Additionally, the silicon wafer suppliers were able to deliver extremely pure silicon slices with resistivity greater than 10000 Ω·cm.

Extensive investigations were performed on the characteristics and behaviour of these extremely pure silicon wafers by applying Si-MBE and X-ray lithography to check whether these techniques are suitable for a SiMMWIC technology.

8.2.2 Influence of Si-MBE and X-Ray Lithography on Characteristics of High Resistivity Silicon

The influence of Si-MBE and X-ray lithography in the fabrication process of SiMMWICs was investigated by measuring the electrical characteristics of the highly insulating substrates, by determining the

145

attenuation of microstrip lines, and by attenuation measurements in waveguides.

The electrical characteristics were measured by the spreading resistance method [8.18] and by the Hall effect. The spreading resistance leads to the well-known method for dopant profiling. Figure 8.10 shows a Ga dopant profile on high-resistivity silicon, grown at 550°C with a preceding thermal cleaning process at 900°C for 5 min. This dopant profile demonstrates the high capability of Si-MBE. A constant doping level of $1 \cdot 10^{18}$ cm^{-3} is found in the MBE layer and an abrupt junction to the high resistivity silicon substrate is observed. The substrate doping process is not influenced by the MBE process. In general, no change of the electrical characteristics was found after various process steps like thermal oxidation, pyrox deposition, MBE, plasma etching, wet chemical etching, cleaning procedures, evaporation of metal film, etc.

The characteristics of microstrip lines were investigated between 90 and 100 GHz using ring resonators and linear microstrip resonators. Measurements on ring resonators yielded an effective dielectric constant of 9.2 for the investigated 10000 Ω·cm silicon substrate. The attenuation of microstrip lines was measured on linear microstrip resonators of different lengths. This was performed on substrates with different conductivities and different treatments. For 10000 Ω·cm substrates, a line attenuation of 0.6 dB/cm was found, for 650 Ω·cm p-silicon the attenuation was about 1.4 dB/cm, and for 100-200 Ω·cm n-substrates the attenuation was more than 2 dB/cm.

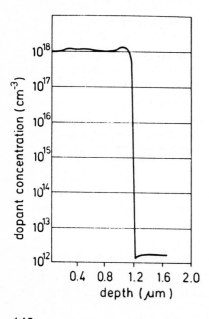

Fig.8.10. Ga dopant profile on a high resistivity silicon wafer grown by MBE at 550°C

In silicon technology, isolation layers play an important role and are frequently used. For this reason microstrip lines on high resistivity silicon with isolation layers were investigated. Thermally deposited SiO_2, 100 nm thick and deposited on both sides of the wafer, and evaporated Al_2O_3, 150 nm thick and deposited on one side of the wafer, were examined. For the microstrip line on the 150 nm thick Al_2O_2 isolation layer, an attenuation of 0.6 dB/cm was found; for microstrip lines on the SiO_2 isolation layers, a slightly increased attenuation of 0.7 dB/cm was measured.

Microstrip lines were also fabricated on MBE processed wafers. For this Ga- and Sb-doped MBE layers with a concentration of $2 \cdot 10^{17}$ cm^{-3} and a thickness of 1 - 2 μm were deposited at 750°C. Microstrip lines were then fabricated, both on wafers with these MBE layers, and on wafers where the MBE layers were etched away. The patterning of the strip lines was done either by photolithography or by X-ray lithography. On substrates where the MBE layer was etched away, the attenuation was found to be 0.6 dB/cm, the same value as on unprocessed wafers; on substrates with a 1.5 μm thick Ga or Sb MBE layer the attenuation was 1.8 dB/cm. In Table 8.2 the measured attenuation values for microstrip lines on silicon substrates are summarized.

Table 8.2. Attenuation values of 50 Ω microstrip lines on different silicon substrates and different additional layers (linewidth: substrate height 200 μm, conductor (gold) thickness: 1.5 μm)

ρ [Ω·cm]	w/h	additional layer	α [dB/cm]
10000	1		0.59
650	1		1.4
100-200 (n-Si)	1		>2
10000	1	150nm Al_2O_3 (evaporated, front side)	0.59
10000	1	100nm SiO_2 (therm., 1000°C, both sides)	0.7
10000	1	1.5 μm MBE-layer (2×10^{17} Sb/cm^3)	1.8
10000	1	MBE-layer etched	0.59

Various 10000 Ω·cm silicon samples were also investigated in a reduced height waveguide using a resonator method in the frequency range 75-110 GHz. The relative permittivity was found to be 11.68 (±0.7%), the loss tangent was $1.3 \cdot 10^{-3}$ (±30%) for unprocessed samples and $1.8 \cdot 10^{-3}$ for a sample with a MBE layer of 0.1 μm thickness and a doping concentration of $2 \cdot 10^{16}$ Sb/cm^3.

X-ray lithography is a promising technique for future high-volume and high-yield fabrication of submicron devices [8.19]. Some of the advantages of X-ray lithography can find useful application in manufacturing SiMMWICs. The high resolution may be used for the fabrication of sensitive receiver diodes, high power ring-like IMPATT diodes, or high frequency amplifiers; the uniform in-depth exposure and high aspect ratio (topographic resolution) may be applied to the fabrication of low capacitance air bridges; and the high edge steepness and high aspect ratio may be used for the definition of low-loss microstrip lines and high quality planar antennas.

In a first study, the effects of X-ray exposure on the high resistivity silicon substrate were investigated. This was done to check whether the X-ray exposure increases the low conductivity of the substrate by possible activation of impurity atoms, thus making the substrate unsuitable for millimeter wave integration (increase of attenuation). Several 10000 Ω·cm substrates were exposed to the BESSY (Berlin Electron Storage Ring for Synchrotron Radiation) spectrum [8.19] with different doses. The resistivity of the exposed samples was determined with the spreading-resistance probe and from Hall measurements, and compared with unexposed samples. Even for an exposure of 100 J/cm^2, which is 100 times the exposure necessary for an insensitive resist (PMMA), no degradation of resistivity takes place. Hall measurements confirm these results.

X-ray lithography has been successfully applied for the fabrication of coplanar Schottky diodes on high resistivity silicon [8.20], for manufacturing a MOSFET-tetrode with 0.4 μm gates [8.21] and permeable base transistors [8.22].

8.2.3 Integrated Oscillator

The most powerful solid-state sources at 100 GHz are silicon IMPATT diodes. Continuous-wave power of up to 1 Watt has been achieved in waveguide resonators for quasi read double drift region diodes made by Si-MBE. To realize monolithic integrated oscillators the active device (IMPATT diode) has to be integrated in a resonator and coupled to the outer world by transmission lines and dc bias networks (Fig. 8.11).

Fig.8.11. Photograph of a hybrid integrated oscillator showing the disk resonator, output transmission line, low-pass filter network and bias pad

Disk resonators are effective for oscillator circuits. This disk resonator can be described as a lossy radial waveguide. The matching of the IMPATT diode to the disk resonator was calculated, and satisfies the condition for oscillator stability [8.23]. To be flexible in optimizing the resonator circuit, the first oscillators were fabricated in a hybrid integrated form [8.24]. The oscillator circuit with disk resonator, bias network and microstrip transmission line was manufactured on 100 μm

149

thick highly insulating silicon substrates (Fig.8.11). To reduce skin-effect attenuation, the oscillator circuit was electroplated with gold to a thickness of 3 μm. As active devices, quasi Read double drift IMPATT diodes made from Si-MBE material are used and mounted in the centre hole of the disk resonator directly on the gold-plated copper carrier. Electrical connection to the oscillator circuit is achieved by crossed ribbons.

The RF power and efficiency as a function of dc bias current is shown in Fig.8.12. The maximum cw output is 200 mW with an efficiency of 4.5% at a frequency of 73 GHz. The low threshold current indicates very low losses in the resonator circuit.

For monolithic millimeter wave IMPATT oscillators two approaches have been used, the Monolithic Microstrip Oscillator (MMO) and the Monolithic Coplanar Oscillator (MCO) [8.25]. A cross-section of the monolithic microstrip oscillator is shown in Fig.8.13. The IMPATT diode is grown by Si-MBE, mesa etched and contacted from the rear of the wafer by anisotropic etching and filling the conical hole by gold electroplating. The gold electroplated via a hole also serves as a heat sink. The upper contact to the monolithically integrated IMPATT diode is made by the resonator circuit. A sharp breakdown in the I-V characteristics is found, but there are still problems with the thermal resistance due to incomplete filling of the heat sink.

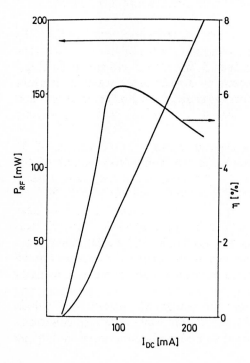

Fig.8.12. RF output power and efficiency of a hybrid integrated Impatt oscillator as a function of dc bias current

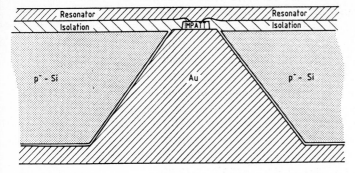

Fig.8.13. Cross section of the monolithic microstrip oscillator with integrated heat sink

For this reason a monolithic coplanar oscillator was investigated. Figure 8.14a shows the layout of the monolithic coplanar oscillator. The disk resonator has been slit in the middle and the integrated IMPATT diode is contacted by the left and right side of the slit disk resonator (Fig.8.14b). Preliminary results are an output power of 1 MW in cw (continuous wave) operation at 76 GHz [8.26].

8.2.4 Monolithically Integrated Transmitters and Receivers

By using IMPATT diodes as the active element, and coplanar Schottky diodes as the nonlinear elements, complete transmitters and receivers, including antennas, can be fabricated in planar silicon technology. Figure 8.15 shows the basic concept of a millimeter wave transmitter link at 100 GHz, the first realized transmitter and receiver chips, and preliminary output power data of the transmitter and a sensitivity curve of the receiver.

The microstrip line antenna used in these circuits consists of 36 radiating elements on an area of 5.4 x 5.6 mm^2. The weight of the elements was calculated after Dolph-Chebyshev. The half-power beam width of the antenna is 23° and the side lobe attenuation is 12 dB.

The nonlinear element in the receiver circuit is a coplanar Schottky diode. In the fabrication process of this diode, Si-MBE was applied for the deposition of the thin (100-200 nm) lightly doped epitaxial layer and X-ray lithography was used for the definition of small Schottky anode fingers. A preliminary result for the receiver sensitivity is 65 μV/μW/cm^2 at 93 GHz.

In conclusion, by combining the advanced methods of silicon molecular beam epitaxy (Si-MBE) and X-ray lithography, a technology for the fabrication of monolithic integrated millimeter wave circuits on

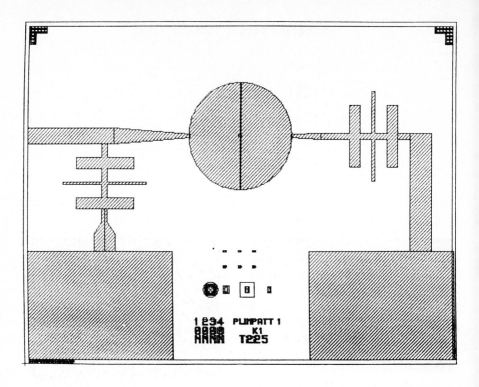

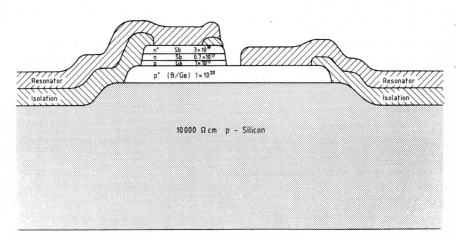

Fig.8.14a,b. (a) Layout of the monolithic coplanar oscillator. (b) Cross section of the Impatt diode integrated in the monolithic coplanar oscillator

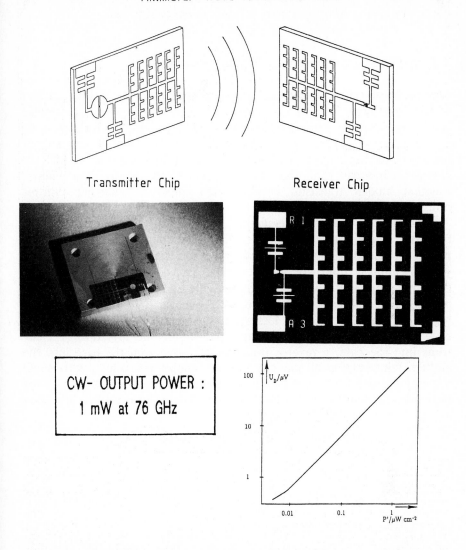

Fig.8.15a-c. Monolithic integrated transmitters and receivers for a millimeter wave transmitter link, **(a)** basic concept, **(b)** the first realized transmitters and receivers on high resistivity silicon **(c)** preliminary power and sensitivity data

high-resistivity silicon has been developed. This technology pushes the upper frequency limit of silicon circuits towards 100 GHz.

8.3 Si/SiGe-Heterojunction Bipolar Transistors

The principles of a Heterojunction Bipolar Transister (HBT) with a wide gap emitter was first proposed by *Shockley* [8.27] in 1951 and considered in more detail by *Kroemer* [8.28, 29] in 1957. The basic idea is to provide an additional energy barrier ΔE_v to holes injected from the base into the n-emitter. Therefore, the emitter material must have a larger band gap than the base material. Consequently, the hetero-junction bipolar transistor is also often called a "wide-gap emitter transistor".

To gain a better understanding of a wide gap emitter bipolar transistor, some basic principles of homojunctions and heterojunctions must be recalled. Figure 8.16 shows the band diagram of a npn homo-junction bipolar transistor under normal bias conditions. The emitter-base junction is forward biased and the collector-base junction is reverse biased. Electrons are injected from the emitter into the base region. J_s and J_r represent losses due to interface recombination and bulk recombination effects, respectively. Beside the injection of electrons into the base, a reverse injection of holes from the base into the emitter takes place. In modern silicon bipolar transistors with small dimensions, the bulk recombination effect can be neglected, and the base current density is clearly dominated by the injection of minority

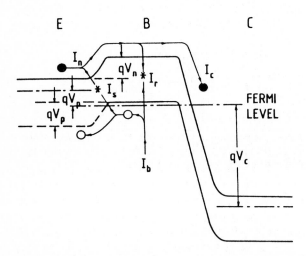

Fig.8.16. Energy band diagram of a homojunction bipolar transistor

154

carriers from the base into the emitter. This injection leads to a reduction in the current gain and to a charge storage problem in the emitter.

An important figure of merit characterizing the emitter-base junction of a transistor is the emitter efficiency γ given by

$$\gamma = \frac{J_n}{J_n+J_p} = \frac{1}{1 + J_p/J_n} \tag{8.1}$$

where J_n and J_p are the densities of the electron and hole currents, respectively, according to Fig.8.16. Since the current gain α for the common base configuration (α_t: base transport factor)

$$\alpha = \gamma \cdot \alpha_t \tag{?.2}$$

depends directly on the emitter efficiency, it is very important that γ is close to unity, meaning that the hole injection current density should be reduced to zero.

In order to obtain a small ratio of hole to electron current for a homojunction, it is necessary to dope the n-emitter side of the junction much more heavily than the p-base side. There are practical limits, however, such as the band-gap narrowing effect.

For a heterojunction, the hole injection is reduced drastically by the additional energy barrier in the valence band. Figure 8.17 shows the band diagram of a heterojunction bipolar transistor under normal bias conditions. It is obvious that the energy barrier qV_n for the elec-

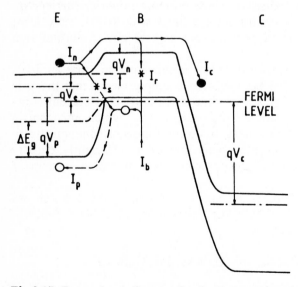

Fig.8.17. Energy band diagram of a double heterojunction bipolar transistor [8.29]

tron injection (J_n) is smaller than the barrier qV_p for the reinjection of holes (J_p). Compared to a homojunction, the reduction of the J_p/J_n ratio can be expressed to first order by the exponential factor:

$$\frac{J_p/J_n \text{ (heterojunction)}}{J_p/J_n \text{ (homojunction)}} \simeq \exp(-\Delta E_g/kT) \qquad (8.3)$$

where ΔE_g is the difference between the emitter and base band gaps and kT = 0.025 eV at room temperature.

Due to the strong influence of the exponential factor, the emitter efficiency of a heterojunction is always close to unity even for a small difference (0.15-0.3 eV) in band gaps and therefore independent of the doping levels in the emitter and the base. This provides a new design flexibility in the choice of doping levels in the emitter and the base.

A high base doping concentration leads to many advantages, such as low base resistance, low noise value and high current capability of the transistor. Base doping concentrations as high as $2 \cdot 10^{20}$ cm^{-3} have been reported so far for GaAs heterojunction bipolar transistors [8.30]. On the other hand, the low emitter doping concentration produces small emitter junction capacitance, and consequently a high cutoff frequency and a high emitter breakdown voltage.

For digital application with saturated logic circuits, such as I^{2}L, the propagation delay is limited by the total charge stored in the emitter of the switching transistor. Since no more minority carriers are injected in the case of a circuit implemented with HBTs, significant improvements of the dynamic performance and the power consumption can be obtained [8.31].

The energy-gap difference does not mean, however, that an arbitrarily high current gain can be obtained. It simply means that the hole injection current I_p becomes a negligible part of the base current compared to the two recombination currents I_s and I_r, which also contribute to the base current. To have a useful transistor, one must still have low recombination currents. This implies a sufficiently defect-free hetero-interface to keep the interface recombination current I_s low.

There are many possible wide-gap/narrow-gap systems which are lattice matched and are suitable for fabrication of a heterojunction bipolar transistor. These include:

GaAs/Ge, GaAlAs/GaAs, InP/GaInAs, InP/GaInAsP

and other III-V material combinations.

The GaAs/Ge heterojunction system was the first to be studied extensively. However, concerning bipolar applications, this material combination has only achieved minor importance. Among the many other combinations of materials used to form an emitter-base hetero-junction, the GaAlAs/GaAs and the InP/GaInAs systems appear to be the most important. Impressive results have been achieved in the GaAlAs/GaAs system. A propagation delay time of as little as 14.2 ps/gate in a ring oscillator and a maximum toggle frequency of up to 20 GHz in a frequency divider were observed [8.32] with HBTs of 1.2 × 9 μm^2 emitter dimensions. The cutoff frequency f_T was about 67 GHz at a current density of $6 \cdot 10^4$ A/cm^2. These high-speed performances were mainly attributed to the heterojunction principle, and only secondarily determined by the specific properties of GaAs.

Today's conventional silicon integrated circuits (Si-IC) are based on homojunction transistors. The high-frequency performance of the silicon homojunction transistors has been constantly improved in the past by the new development of spacer and self-aligned technologies. The implementation of a heterojunction to the silicon material system would further extend the existing high-speed performance of today's Si bipolar technology, and would be of great importance for future analogue and digital applications.

It is an attractive idea to use a wide-gap material for replacing the Si homojunction emitter in order to fabricate a Si-based hetero-junction bipolar transistor. Different approaches have been reported in the past by several groups. One approach is to use amorphous silicon or microcrystalline silicon as an emitter, which has a large band gap ($\simeq 1.7 \text{eV}$), creating a real heterojunction with the crystalline silicon base [8.33]. Drawbacks of this approach are the high emitter resistance, the interface state density, and the difficulty in obtaining low contact resistances on this noncrystalline material. Another solution, which is sometimes called a pseudo-heterojunction, utilizes the presence of a thin interface oxide between the base and the polysilicon emitter. The back injection of holes from the base is now limited by tunnelling through the thin oxide barrier [8.34]. For this reason, a precise control of the oxide layer thickness is needed.

The other Si-based heterosystem of interest is the monocrystalline Si/SiGe system. Figure 8.18 shows the lattice constant and the band gap for the well-established III-V systems and the new Si/SiGe material combination. A characteristic feature of the Si/SiGe heterosystem is the large difference of the lattice constant (4% for germanium and silicon).

This mismatch produces strain, which strongly influences the band structure. When SiGe is grown on a Si substrate (Si is unstrained, SiGe is fully compressed), the whole difference in the band energy is avail-

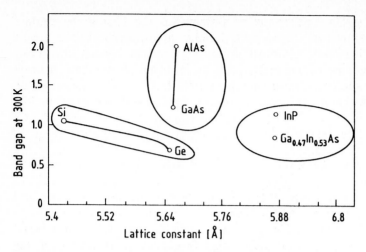

Fig.8.18. Lattice constant and band gap for different heterosystems

able as difference in the valence band ΔE_v (Fig.8.19), resulting in the ideal situation for a hetero-bipolar junction. Figure 8.20 shows the band gap of an unstrained SiGe alloy and of SiGe grown onto a thick Si substrate. A germanium molar fraction of 0.2 would result in a difference in the band gap of about 0.2 eV [8.35] which is quite sufficient for a heterojunction since it improves the J_n/J_p ratio of the emitter by a factor of about 3000.

To have a useful heterojuction in terms of low interface state densities, misfit dislocations have to be avoided. For thin SiGe films up to the critical thickness t_c, elastic accommodation prevents the formation of misfit dislocation (Fig.8.21). Above the critical thickness, misfit dislocations are generated thus diminishing the build-up of strain (Fig.8.21 right).

Figure 8.22 shows the critical thickness t_c versus lattice mismatch. The experimental results indicate a strong dependence upon the growth temperature [8.36]. In order to have a high critical thickness, the temperature during epitaxial growth of the layer sequence should be as low as possible, and no subsequent fabrication process should exceed this temperature.

Key issues for the technological realization of a heterojunction bipolar layer sequence are:

Fig.8.19. Band alignment of strained SiGe on unstrained Si substrate

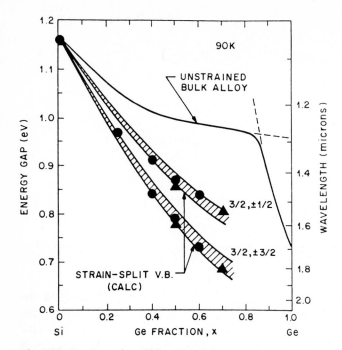

Fig.8.20. Band gap in a SiGe alloy [8.35]

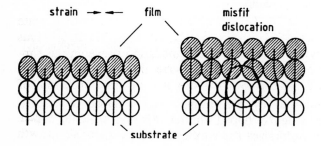

Fig.8.21. Mismatch accommodation by strain (left) or misfit dislocations (right)

- low growth temperature (avoids interdiffusions and increases the critical thickness of the base layer),
- precise control of layer compositions and thickness,
- crystal perfection of the layer,
- high level of dopant incorporation.

MBE is ideally suited for meeting the above requirements.

An example of a Si/SiGe heterojunction bipolar layer sequence is given in Fig.8.23. Starting growth with a silicon substrate, the n-Si

5000
1000
100
10
1
t_c (nm)
EXPERIMENT
MBE-750°C
MBE - 550°C
THEORY
X = 0.25
$Si_{1-X}Ge_X/Si$
0.001
0.01
η

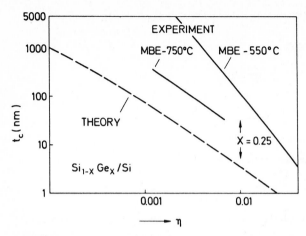

Fig.8.22. Critical thickness t_c versus mismatch n. Comparison of theory with MBE experiments [8.36]

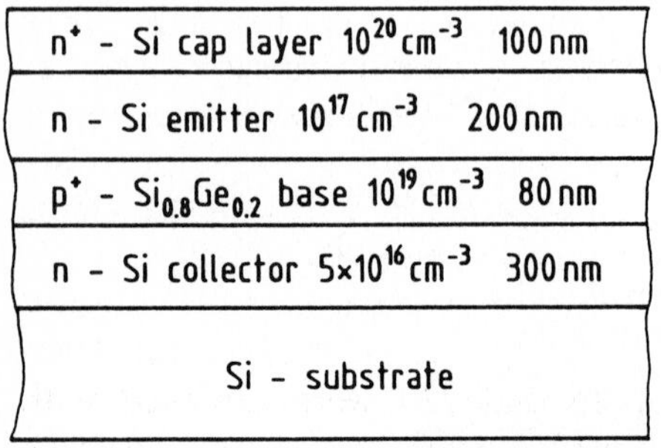

n^+ - Si cap layer 10^{20} cm^{-3} 100 nm
n - Si emitter 10^{17} cm^{-3} 200 nm
p^+ - $Si_{0.8}Ge_{0.2}$ base 10^{19} cm^{-3} 80 nm
n - Si collector 5×10^{16} cm^{-3} 300 nm
Si - substrate

Fig.8.23. Si/Ge heterojunction bipolar transistor layer sequence

collector is grown, followed by the strained highly doped SiGe base layer and the unstrained lightly doped emitter. Finally, a highly doped Si cap layer on the top minimizes the emitter contact resistance.

The main problem in the fabrication process is the contacting of the thin (typically 80nm) base layer. Three ways have been reported so far.

Selective etching of Si against SiGe is one possibility for making contact to the base [8.37]. The etchant must be strain selective with a high selectivity coefficient, and the etched surface should be smooth. After exposing the base layer, the emitter and the base contacts are evaporated. The collector contact is made from the rear. Finally, device isolation is performed by mesa etching. Figure 8.24 shows a cross-section of a mesa transistor.

Ion implantation is an alternative method for base contacting. This technique was successfully applied in conjunction with a nonselective

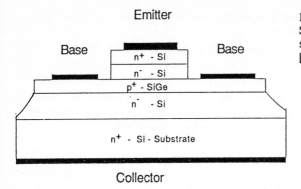

Emitter

Base | | Base

| n+ - Si |
| n⁻ - Si |
| p+ - SiGe |
| n⁻ - Si |
| n+ - Si - Substrate |

Collector

Fig.8.24. Cross section of a Si/SiGe hetero-bipolar transistor with mesa isolation [8.33]

dry etching step for the first reported Si/SiGe HBT [8.38]. One of the critical steps is the annealing of the p-type extrinsic base implants. High electrical activation of the implant tends to require high-temperature annealing. On the other hand, it is important to minimize diffusion of the p-base dopant into the emitter. In particular, if the dopant B or Ga diffuses significantly into the emitter, the emitter-base junction becomes a homojunction with a poor emitter efficiency. Another limitation is the maximum allowed temperature at which relaxation of the strain by generation of misfit dislocations occurs. An appropriate solution is the rapid thermal annealing process, in which Ga implants are activated in heating cycles of 20 s duration at 500°C [8.39].

An NEC group [8.40] published a collector top design and a novel realization using a two-step differential epitaxy to contact the base. Room temperature current gains of 15 and 250 were achieved with $Si_{0.7}Ge_{0.3}$ base layers of $5 \cdot 10^{19}$ and $8 \cdot 10^{17}$ cm^{-3} base doping, respectively. These results can be compared to current gains of 1 and 100, respectively, for Si homojunction transistors of the same size and doping levels.

All the published results exhibit the expected increase in emitter efficiency with lower band gap in the base. The Si/SiGe heterosystem has been shown to be a particularly interesting material. It will clearly extend the present potential of the homo-silicon-based devices, and will add more flexibility to the device design and enable the development of completely new structures for future applications.

8.4 Conclusions

For the dynamics of microelectrons, the interplay of technology, device design and system concepts is essential. The new method of Si-MBE offers many innovative technological inputs for realization of existing device types with improved properties and for tailoring of novel materials and circuit structures.

Industrial equipment is indeed not available at the moment, but very positive developments towards such industrial equipment stem from cooperation between the electronics industry and equipment manufacturers. These authors have no doubt that industrial equipment will become available, and the more device designers and system engineers press for new innovations, the earlier this will be.

Acknowledgement. Help from the Si-MBE group and discussions with colleagues at the AEG Research Center, Telefunken Electronics, Ruhr University Bochum (Prof. Bosch), and Technical University of Munich (Prof. Russer) is acknowledged.

References

8.1 Silicon Molecular Beam Epitaxy, ed. by E. Kasper, J.C. Bean (CRC, Boca Raton 1988)
M.A. Herman, H. Sitter: *Molecular Beam Epitaxy*, Springer Ser. Mat. Sci., Vol.7 (Springer, Berlin, Heidelberg 1989)
8.2 H. Jorke: Presented at MBE-V, Sapporo (1988)
8.3 E. Kasper: Appl. Phys. A **28**, 129 (1982)
K. Voigtländer, H. Risken, E. Kasper: Appl. Phys. A **39**, 31 (1986)
8.4 G.J. Fisanick, H.J. Gossmann, P. Kuo: MRS Symp. Proc. Vol.102, ed. by R.T. Tung, L.R. Dawson, R.L. Gunshor (MRS, Pennington 1988) p.25
8.5 L. Vescan, E. Kasper, O. Meyer, M. Maier: J. Crystal Growth **73**, 482 (1985)
8.6 E. Kasper, K. Wörner: J. Electrochem. Soc. **132**, 2481 (1985)
8.7 H.J. Herzog: Esprit project no. 305, to be published
8.8 E. Kasper, H.J. Herzog, K. Wörner: J. Crystal Growth **81**, 458 (1987)
8.9 U. König: to be published
8.10 E. Kasper: *Advances in Solid State Physics* **27**, 265 (Vieweg, Braunschweig 1987) p.265
8.11 R.K. Hoffmann: *Integrated Microwave Circuits* (Springer, Berlin, Heidleberg 1986)
8.12 D.M. McQuiddy, J.W. Wassel et al: IEEE Trans. MTT-**32**, 997 (1984)
8.13 T.M. Hyltin: IEEE Trans. MTT-**13**, 777 (1965)
8.14 B.W. Battershall, S.D. Emmons: IEEE Trans. SC-**3**, 107 (1968)
8.15 A. Ertel: Electronics **76**, (Jan. 1967)
8.16 A. Rosen, et al.: RCA Rev. **42**, 633 (1981)
8.17 K.M. Strohm, J. Hersener, H.J. Herzog: Proc. Eurocon '86 (Paris, 1986) AI.4
8.18 R. Brennon: Solid State Technol. 127 (Dec. 1984)
8.19 A. Heuberger: Microelectronic Engineering **3**, 535 (1985)
8.20 K.M. Strohm, J. Hersener, E. Piper: Microcircuit Engineering, Wien (1988) to be published
8.21 J. Hersener, E. Piper, A. Wilhelm, G. Birkenstock: Microcircuit Engineering, Wien (1988) to be published
8.22 J. Pustai: Microwaves & RF 173 (March 1987)
8.23 J. Buechler, E. Kasper, J.F. Luy, P. Russer, K.M. Strohm: European Microwave Conf., Stockholm (1988) to be published
8.24 J. Buechler, E. Kasper, J.F. Luy, P. Russer, K.M. Strohm: Electron. Lett. **24**, 977 (1988)

8.25 J.F. Luy, K.M. Strohm, J. Buechler: European Microwave Conf., Stockholm (1988) to be published
8.26 J.F. Luy, K.M. Strohm, J. Buechler: Microw. Opt. Techn. Lett. 1, 117 (1988)
8.27 W. Shockley: Circuit element utilizing semiconductive material. U.S. patent 2569347 (1951)
8.28 H. Kroemer: Proc. IRE 45, 1535 (1957)
8.29 H. Kroemer: Proc. IEEE 70, 13 (1982)
8.30 J.L. Lievin, C. Dubon-Chevallier, F. Alexandre, G. Leroux, J. Dungle, D. Ankri: IEEE EDL-7, 129 (1986)
8.31 P. Nerozny, H. Beneking, R.J. Fisher, H. Morkoc: IEEE Trans. ED-33, 1238 (1986)
8.32 K.C. Wang, P.M. Asbeck, M.F. Chang, G.J. Sullivan, D.L. Miller: IEEE EDL-8, 383 (1987)
8.33 J. Symons, M. Ghannam, J. Nijs, A. van Ammel, P. de Scheper, A. Neugroschel, R. Mertens: Proc. U.S.-Belgium Joint Seminar (1986) Leuven, Belgium
8.34 M.A. Green, R.D. Godfrey: IEEE EDL-4, 225 (1983)
8.35 R. People: Phys. Rev. 32, 1405 (1985)
8.36 E. Kasper, H.J. Herzog, H. Dämbkes, Th. Ricker: In *Two-Dimensional Systems: Physics and New Devices*, ed. by G. Bauer, F. Kuchar, H. Heinrich, Springer Ser. Solid-State Sci. Vol.67 (Springer, Berlin, Heidelberg 1986) p.52
8.37 P. Narozny, M. Hamacher, H. Dämbkes, H. Kibbel, E. Kasper: To be published in IEDM Technical Digest 1988
8.38 G.L. Patton, S.S. Iyer, S.L. Delage, S. Tiwari, J.M. Stork: IEEE EDL-4, 165 (1988)
8.39 H.B. Harrison, S.S. Iyer, G.A. Sai-Hulasz, S.A. Cohen: Appl. Phys. Lett. 13, 992 (1987)
8.40 T. Tatsumi, H. Hirayama, N. Aizaki: Appl. Phys. Lett. 52, 895 (1988)

Subject Index

Heteroepitaxy 141,142
Heterojunction bipolar transistor
   14-17,118-131
- Si/SiGe 154-161
High-doping effect 19
Hydrogenated
- amorphous silicon (a-Si:H)
   121-123
- amorphous silicon carbide
   (a-SiC:H) 123-125
- micro-crystalline silicon
   (μc-Si:H) 125,126

IIL (integrated-injection logic)
   circuits 101-103
IMPATT (impact ionization
   avalanche transit time) 148-151
ISAC (implanted self-aligned
   contact) 80-93
Isolation
- leakage 74
- width 66

Laser doping 36
Lattice mismatch 126-128,
   157,158
Link resistance 19
LOCOS (local oxidation of
   silicon) technique 61
Low energy implantation 50,
   51
Low processing temperature of
   MBE 136,137
LSS (Lindhard-Scharff-Schiott)
   theory 31

Macrocell array (MCA) 89
Microstrip line 143
Modelling problems 17-20
Molecular beam epitaxy (MBE)
   135-162
- Si 135-162

NTL (non-threshold logic) 20

Oscillator circuits 148-151
Oxide layer
- native 7,44
- thickness 7,49

Packing density 62,70
Planarization process 65
PMMA (polymethyl-metha-
   crylate) 148
Power
- consumption 6
- dissipation 86,102
Preamorphization 32,33,51,52
Propagation delay time 80-87,95

RAM (random access memory) 3
RF (radio frequency) power 125
RIE (reactive ion etching) 62,63
ROA (rapid optical annealing) 46
RTA (rapid thermal annealing) 29

SBC (standard buried collector) 4
Scaling problems 11-14
SCOT (salicide self-aligned
   silicide) base contact technology
   79-88
SDD (self-aligned double diffusion polysilicon technology) 71
Self-aligning
- process 5,9-11
- technology 29
SEM (scanning electron microscopy) 64
SEPOX (selective polysilicon
   oxidation) 74
Series resistance 19
Shadowing effect 32
SICOS (sidewall base contact
   structure) 14,95-110
Sidewall
- effect 18
- region 13,14
- spacer 37,38
Silicide 5,82,83

Silicon monolithic milimeter wave
  integrated circuits (SiMMWICs)
  142-154
SIMS (secondary-ion-mass spec-
  troscopy) 34
SIPOS (semi-insulating
  polycrystalline silicon) 121
SST (super self-aligned process
  technology) 96
Surface
- contamination 136,137
- recombination velocity 9,44-46
Symmetrical structure 101

Transit frequency (*see* Cut-off
  frequency)
Transit time 18,54-58,80-82,
  106,107
Trench

- etching 62,63
- filling 64,67,70
- isolation 61-74
Tunneling
- current 7,8
- model 47-49

Upward/downward operation
  100,101

VSC (variable size cell) masterslice 89

WKB (Wentzel, Kramer, and
  Brillouin) approximation 48

X-ray lithography 145-148
XTEM (cross-sectional transmission electron microscopy)
  47-51